# Stellar Theology and Masonic Astronomy

## *Robert Brown*

Must Have Books
503 Deerfield Place
Victoria, BC
V9B 6G5
Canada


ISBN: 9781773237831

Copyright 2021 – Must Have Books

# CONTENTS

## *Part First*

# Part Second

*Arranged in the Form of a Masonic Lecture,
and illustrated by a Zodiacal Diagram*

# Part Third

# Chapter 8

## Astronomical Explanation (Continued) ............................ 157

# LIST OF ILLUSTRATIONS

# Stellar Theology and Masonic Astronomy

## Part First

**Chapter 1**

# INTRODUCTION— A FEW WORDS TO THE MASONIC FRATERNITY

THE WRITER OF THIS WORK was for a long time in considerable doubt as to the propriety of its publication—not because he had any lack of faith in the truth of the theory it advocates, but from a fear that the revelations it contains might be thought unlawful according to a strict construction of the masonic obligation. But, after consulting with many conscientious as well as eminent members of the fraternity, the author was confirmed in his belief that nothing is said in the book which discloses any of the *essential secrets* of that order.

The "essential secrets" of freemasonry are defined by Dr. Oliver, in his "Dictionary of Symbolical Masonry," as consisting of nothing more,

> than the signs, grips, passwords, and tokens essential to the preservation of the society from the inroads of impostors, together with certain symbolical emblems, the technical terms apper-

taining to which serve as a sort of universal language by which the members of the fraternity can distinguish each other in all places and countries where lodges are instituted.

Now, although in the following pages the masonic tradition as to the history of an important masonic personage is freely alluded to, nowhere is there anything said, or even implied by which any of the essential secrets of the craft are placed in peril; nor is there any particle of information given which can be use to unprincipled persons, however acute, who might desire to impose themselves upon the fraternity as having a right to its benefits and honors. The masonic reader should also bear in mind that many things in the following pages, which are to him full of masonic significance, will appear to the uninitiated but an expression of some of the simplest facts in the science of astronomy, long established and known to all.

Says Gadicke, a masonic writer of repute:

> With the increase of enlightenment and rational reflection, it is admitted that a brother may both speak and write much upon the order without becoming a traitor to its secrets.... Inquiries into the history of the order, and *the true meaning of its hieroglyphics and ceremonies* by learned brethren, cannot be considered treason, for the order itself recommends the study of its history, and that every brother should instruct his fellows as much as possible. It is the same with the printed explanation of the moral principles and *symbols* of the order. We are recommended to study them incessantly, until we have made ourselves masters of the valuable information they contain; and, when our learned and cautious brethren publish the result of their inquiries, they ought to be most welcome to the craft.

These remarks of Gadicke are quoted with approbation by Dr. Oliver, who himself says, in the introduction to his "Landmarks":

No hypothesis can be more untenable than that which forebodes evil to the masonic institution from the publication of scientific treatises illustrative of its philosophy and moral tendency. The lodge lectures, in their most ample and extended form, however pleasing and instructive soever they may be, are unsatisfactory and inconclusive. They are merely elementary, and do not amply and completely illustrate any one peculiar doctrine. As they are usually delivered in nine tenths of the lodges, they are monotonous, and not perfectly adapted to the end for which they are framed, or for the effect they are intended to produce. For this reason it is that literary and scientific men, who have been tempted to join our ranks in the hope of opening a new source of intellectual enjoyment, and of receiving an accession of novel ideas for their reflection and delight, so frequently retire, if not with disgust, at least with mixed feelings of sorrow and regret, at the unprofitable sacrifice of so much valuable time which might have been applied to a better purpose.

He adds that,

if the authorized lectures of masonry were amplified and illustrated, such instances would not only rarely occur, but our lodges would become the resort of all the talent and intelligence in the country.

Dr. Mackey, who in America holds the highest rank as a masonic writer, says:

The European masons are far more liberal in their views of the obligation of secrecy than the English or Americans. There are few things, indeed, which a French or German masonic writer will refuse to discuss with the utmost frankness. It is now beginning to be generally admitted—and English and American writers are acting on the admission—that the only real *aporrheta* (essential secrets) of freemasonry are the modes of recognition and the peculiar and distinctive ceremonies of the order, and *to these last* it is claimed that reference may be *publicly made* for the purpose of *scientific investigation,* provided that the reference be made so as to be

obscure to the profane and intelligible only to the initiated.          (Symbolism—Synoptical Index, *Aporrheta)*

Many masons who do not make themselves familiar with the standard and authorized masonic authors, like Dr. Oliver in England, and Pike, Mackey, and Morris in America, are not aware how freely many parts of our ritual are spoken of by brothers occupying the most distinguished positions in the fraternity.

In this work "I have been scrupulously careful about the admission of a single sentence from the peculiar lectures of masonry which has not already appeared in the printed form in one or other of our legitimate publications."

In speaking of the masonic traditions and legends, I have used no greater freedom than other masonic writers whose works are authorized by the highest masonic bodies in England, Germany, France and America; and, in view of all these considerations, have come to the conclusion that it was not wise to permit an unnecessary and unrequired degree of caution to longer delay the publication of truths which are, as I am persuaded, of great importance and interest to the craft.

# Chapter 2

# THE ANCIENT
# MYSTERIES
# DESCRIBED

IF WE CLOSELY EXAMINE the elder forms of religious worship, we will find in most of them that God is worshiped under the symbol of the sun. This is not only true of those nations called pagan, but we also find in the Bible itself the sun alluded to as the most perfect and appropriate symbol of the creator. The sun is the most splendid and glorious object in nature. The regularity of its course knows no change. It is "the same yesterday, today, and forever." It is the physical and magnetic source of all life and motion. Its light is a type of eternal truth; its warmth of universal benevolence. It is therefore not strange that man in all ages has selected the sun as the highest and most perfect emblem of God. There is a natural tendency, however, in the human mind, to confound all symbols with the person or thing which they were at first only intended to illustrate. In the course of time we therefore find that most nations forgot the worship of the true God, and began to adore the sun itself, which they thus deified and personified. The sun thus personified was made the theme of allegorical history, emblematic of his yearly passage through the twelve constellations.

The zodiac is the apparent path of the sun among the stars. It was divided by the ancients into twelve equal parts, composed of the clusters of stars, named after "living creatures," typical of the twelve months. This glittering belt of stars was therefore called the *zodiac,* that word meaning "living creatures," being derived from the greek word *zodiakos,* which comes from *zo-on,* an animal. This latter word is compounded directly from the primitive Egyptian radicals, *zo,* life, and *on,* a being.

The sun, as he pursued his wan among these "living creatures" of the zodiac, was said, in allegorical language, either to assume the nature of or to triumph over the sign he entered. The sun thus became a Bull in *Taurus,* and was worshiped as such by the Egyptians under the name of Apis, and by the Assyrians as Bel, Baal, or Bul. In *Leo* the sun became a Lion-slayer, Hercules, and an Archer in *Sagittarius.* In *Pisces,* the Fishes—he was a fish—Dagon, or Vishnu, the fish-god of the Philistines and Hindoos. When the sun enters *Capricornus* he reaches his lowest southern declination; afterward as he emerges from that sign the days become longer, and the Sun grows rapidly in light and heat; hence we are told in the mythology that the Sun, or Jupiter, was suckled by a goat. The story of the twelve labors of Hercules is but an allegory of the passage of the sun through the twelve signs of the zodiac, and past the constellations of proximity thereto.

The beautiful virgin of the zodiac, *Virgo,* together with the Moon, under a score of different names, furnishes the female element in these mythological stories, the wonderful adventures of the gods. These fables are most of them absurd enough if understood as real histories, but, the allegorical key being given, many of them are found to contain profound and sublime astronomical truths. This key was religiously kept secret by the priests and philosophers, and was only imparted to those who were initiated into the MYSTERIES. The profane and vulgar crowd were kept in darkness, and believed in and worshiped a real Hercules or Jupiter, whom they thought actually lived and performed all the exploits and underwent all the transformations of the mythology.

By these means the priests of Egypt ruled the people with a despotic power. The fables of the mythology disclosed to them grand scientific truths, and to them only. The very stories themselves served to perpetuate those truths for the benefit of the initiated, and also formed an easy vehicle for their transmission. Books were not only rare and difficult of multiplication, but it is also probable that, in order that scientific knowledge might be concealed, it was considered unlawful to commit it to writing. If in special cases it became an absolute necessity to do so, the sacred hieroglyphs were employed. These were known only to the initiated; there was another sort of written characters used by the common people. (Rawlinson's "Herodotus," Appendix to Book II, Chapter V.)

Science was thus for the most part orally transmitted from one hierophant to another. While an abstruse and difficult lecture is not easy, either to remember or to repeat, on the contrary, a mythological tale can with ease be retained in the memory and communicated to another, together with the key for interpretation. These fables, therefore, served a threefold purpose:

1. They kept the secrets of science from all but those who understood the key to them;

2. Being themselves easy to remember, they served on the principle of the art of *mnemonics,* or artificial memory, to keep alive the recollection of scientific facts which otherwise might be lost;

3. Being the means of keeping the people in ignorance, by their use the priests were enabled to rule them through their superior power of working apparent miracles.

The science in which the Egyptian priesthood were most proficient, and which they most jealously guarded, was that of astronomy. The people worshiped the sun, moon, and stars as gods, and a knowledge of their true nature would have at once put an end to the influence of the priests, who were believed by the ignorant and superstitious crowd to be able to withhold or dispense, by prayers, invocations, and sacrifices, the divine favor. The priest of a pretended god, when once his god is exposed, stands before the world a convicted impostor. To deny the divinity of the sun, moon, and stars, or, what was the same thing, to permit science to disclose their true nature to the masses of the people was consequently held by the priesthood of Egypt as the highest of crimes. By a knowledge of astronomy the priests were able to calculate and to predict eclipses of the sun and moon, events beheld with superstitious awe and fear by the multitude. Seeing how certainly these predictions, when thus made, were fulfilled, the priests were credited with the power to foretell other events, and to look into the future generally. So they cast horoscopes and assumed to be prophets.

Of course, a knowledge of astronomy diffused among the people would have been fatal to these pretensions. The facts of astronomy were therefore, for these reasons most carefully hidden from the common people, and the priesthood only communicated them to each other, veiled in allegorical fables, the key to which was disclosed to him only who had taken the highest degrees of the Mysteries, and given the most convincing proofs of his fidelity and zeal.

The names under which the sun was personified were many, but the one great feature, most prolific of fables, was his great decline in light and heat during the winter, and his renewal in glory and power at the vernal equinox and summer solstice, which gave rise to all that class of legends which represent the sun-god (under various names) as dying and being restored to life again.

Thus, we are told, in the Egyptian sacred legend, that *Osirus,* or the Sun, was slain by *Typhon,* a gigantic monster, typical of darkness and the evil powers of nature. The body was placed in a chest, thrown into the Nile, and swept out to sea. Isis, or the Moon personified as a goddess, ransacks the whole earth in search of the body, which she finds horribly mutilated. She joins the dissevered parts, and raises him to life again.

In the Greek mythology we are told that *Adonis* (the Lord, or sun-god) is slain, but it returns to life again for six months each year—thus dying in the fall and winter months and returning to life again during the spring and summer.

The ritual of the Mysteries in Egypt, India and Greece, was founded upon this legend, in some form, of the death and resurrection of the personified sun-god.

The Egyptian Mysteries of Osiris and Isis were in the form of a mystic drama, representing the death by violence of Osiris (the sun-god), the search for his body by Isis, the Moon, and its finding and being raised to life and power again. In the celebration of these Mysteries the *neophyte* was made to perform all the mysterious wanderings of the goddess amid the most frightful scenes. He was guided by one of the initiated, who wore a mask representing a dog's head, in allusion to the bright star *Sothis* (Sirius, or the dog-star), so called because the rising of that star each year above the horizon just before day gave warning of the approaching inundation of the Nile. The word *Sothis* means the "barker," or "monitor."

The candidate was by this guide conducted through a dark and mysterious labyrinth. With much pain he struggled through involved paths, over horrid chasms, in darkness and terror. At length he arrived at a stream of water, which he was directed to pass. Suddenly, however, he was assaulted and arrested by three men, disguised in grotesque forms, who taking a cup of water from the stream, forced the terrified candidate to first drink of it. This was the water of forgetfulness, by drinking which all his former crimes were to be forgotten, and his mind prepared to receive new instructions of virtue and truth.

The attack of *Typhon,* or the spirit of darkness, typical of the evil powers of nature, upon *Osiris,* who was slain, was also enacted as the initiation progressed, and amid the most terrible scenes, during which the "judgment of the dead" was also represented, and the punishments of the wicked exhibited as realities to the candidate. The search for the body of Osiris, which was concealed in the mysterious chest or "ark," followed. The mutilated remains were at last found, and deposited amid loud cries of sorrow and despair. The initiation closed with the return of Osirus to life and power. The candidate now beheld, amid effulgent beams of light, the joyful mansions of the blessed, and the resplendent plains of paradise.

> I saw the sun at midnight" (says Apuleius, speaking of his own initiation into the Mysteries of Isis) "shining with its brilliant light, and I approached the presence of the gods beneath, and the gods of heaven, and stood near and worshiped them.　　　　(See "Metamorphosis")

At this stage of the initiation, all was life, light, and joy. The candidate was himself figuratively considered to have risen to a new and more perfect life. The past was dead, with all its crimes and unhappiness. Henceforth the candidate was under the special protection of Isis, to whose service he dedicated his new life. (See Apuleius.)

The sublime mysteries of religion and the profoundest teachings of science were now revealed to him, and satisfied his thirst for knowledge, while the possession of power as one of the hierarchy gratified his ambition.

The Mysteries of all the other nations of antiquity were quite similar to those of Egypt, and were no doubt derived from them.

In India the chief deity was triune, and consisted of *Brahma,* the Creator, *Vishnu,* the Preserver, and *Siva,* the Destroyer. Brahma was the representative of the rising sun, and the others respectively of the meridian and the setting sun. The aspirant having been sprinkled with water and divested of his shoes, was causing to circumambulate the altar three times.

At the east, west, and south points of the mystic circle were stationed triangularly the three representatives of the sun-god, denoting the rising, setting, and meridian sun. Each time the aspirant arrived in the sough he was made to exclaim, "I copy the example of the sun, and follow his benevolent course."

After further ceremonies, consisting in the main of solemn admonitions by the chief Brahman to lead a life of purity and holiness, the aspirant was again placed in charge of his conductor, and enjoined to maintain strict silence under the severest penalty; told to summon up all his fortitude and betray no symptoms of cowardice.

Amid the gloom then began bewailings for the loss of the sun-god *Sita,* followed by ceremonies of fearful import, and scenic representations of a terrible nature. The candidate was made to personify *Vishnu,* and engaged in a contest with the powers of darkness, which, as the representative of the god, he subdued. This was followed by a dazzling display of light, and a view of Brahma exalted, glorified, and triumphant.

In Persia the candidate was prepared by numerous lustrations performed with water, fire, and honey. A prolonged fast for fifty days in a gloomy cavern followed, where in solitude he endured cold, hunger, and stripes. After this the candidate was introduced for initiation into another cavern, where he was received on the point of a sword presented to, and slightly wounding, his naked left breast. He was next crowned with olive, anointed with the sacred oil, and clad in enchanted armor. He was then taken through the *seven* stages of his initiation. As he traversed the circuitous mazes of the gloomy cavern his fortitude was tried by fire and water, and by apparent combats with wild beasts and hideous forms, typical of the evil powers of nature, in the midst of darkness, relieved only by flashes of lightening and the pealing of thunder. He was next made to behold the torments of the wicked in Hades. This was followed by a view of Elysium, and the initiation concluded by a display of divine light and the final triumph of *Ormuzd,* the sun-god, over all the powers of darkness.

In Greece the Mysteries were denominated as the lesser and greater Mysteries. A chosen few only were admitted to the latter, and they were bound to secrecy by the most frightful oaths.

The Eleusinian Mysteries were performed by the Athenians at Eleusis every fifth year, and were subsequently introduced at Rome by Adrian. These Mysteries were the same as those of Orpheus. A magnificent temple of vast extent having been erected for their celebration at Eleusis, they subsequently became known as the Eleusinian Mysteries. The principal officers who conducted the ceremonies were the Hierophant, the Torch-Bearer, the Priest, the Archon, or King, and the Mystagogue.

The hierophant appeared seated upon a magnificent throne, adorned with gold. He was dressed in a royal robe; over his head a rainbow was arched, and there also the moon and seven stars were seen. Around his neck was suspended a golden globe. These expressive symbols all point all the fact that the hierophant represented the sun. Before him were twenty-four attendants, clad in white robes and wearing golden crowns. These represented the twenty-four ancient constellations of the upper hemisphere. Around him burned with dazzling radiance seven lights, denoting the seven planets. The torch-bearer, whose duty it was to lead the procession when the wanderings of Rhea commenced in search of the body of the lost god, may have been intended to represent the feebler light of the moon, since Rhea and Ceres were both identical, according to Herodotus, with the Egyptian Isis. The duty of the mystagogue was to impose silence on the assembly, and command the profane to withdraw. The priest officiated at the altar, and bore the symbol of the moon, being, like the Egyptian priests of Isis, devoted to her service.

The archon, or king, preserved order, offered also prayers and sacrifices, compelled all unworthy and uninitiated persons to retire at the order of the mystagogue, and punished all who presumed to disturb the sacred rites. The aspirant was required to pass through a period of probation, during which he prepared himself by chastity, fasting, prayer, and penitence. He was then dressed in sacred garments, crowned with myrtle, and blindfolded. After being thus "duly and truly prepared" he was delivered over to the Mystagogue, who began the initiation by the prescribed proclamation:

> *"Exas, exas, este Bebeloi!"* — ("Depart hence, all ye profane!")

> The aspirant was then conducted on a long and painful pilgrimage through many dark and circuitous passages: sometimes it seemed to him as if he were ascending steep hills, walking over flinty ground, which tore his feet at every step, and again down deep valleys and through dense and difficult forests. Meanwhile as he advanced, sounds of terror surrounded him, and he heard the fierce roar of wild beasts and the hissing of serpents. At length, the bandage being removed from his eyes, he found himself in what seem a wild and uncultivated country. The light of day never penetrated this gloomy region, and the pale and spectral glare just served to light up the horrors of the scene. Lions, tigers, hyenas and venomous serpents menaced him at every point while thunder, lightening, fire and water, tempest and earthquake, threatened the destruction of the entire world. He hardly recovers from his surprise and terror,

OSIRIS

THE GOD REPRESENTED EMBLEMATI-
CALLY AS A MAN WITH A BULL'S
HEAD, HIEROGLYPHICALLY DENOTING
THE SUN IN TAURUS. IN ONE HAND HE
HOLDS THE SYMBOL OF ETERNAL
LIFE, IN THE OTHER THE EMBLEM OF
POWER, ABOVE WHICH APPEARS
THE NAME OF THE GOD IN HIERO-
GLYPHICS, WHICH, BY A SINGLE
COINCIDENCE, IS COMPOSED ALMOST
ENTIRELY OF MASONIC EMBLEMS.

his eyes no sooner become accustomed to the twilight of the place, than he discovers before him a huge iron door, on which is this inscription: 'He who would attain to the highest and most perfect state, and rise to the sphere of absolute bliss, must be purified by *fire, air, and water.'* He had scarcely read these words when the door turned on its hinges, and he was thrust into a vast apartment also shrouded in gloom. (Arnold)

Then began the wanderings of Rhea in search of the remains of Bacchus, her body begirt with a serpent, and a flaming torch in her hand, uttering as she goes wild and frantic shrieks and lamentations for her loss. Those already initiated join in, and mix their howlings with hers, blended with mournful music. By means of a certain mechanical contriv-ances (see Salverti's "Philosophy of Magic," vol. i, Chapter X; also, Brewster's "Natural Magic") the plains of Tartarus were presented as realities before his eyes. He beheld the flames amid which the wicked suffered the purification by *fire.* Behind him yawned a dismal and dark abyss, from which issued a burning wind and voices of woe and suffering. Approaching the brink he looks down, and sees some suspended on the sharp points of the rocks, and others impaled on a mighty wheel, which turned without ceasing, thus working their way toward heaven through the purgatorial *air.* The purification by *water* was represented by the horrors of a gloomy lake, into which souls less guilty were plunged. Apuleius also alludes to this purification by fire, air, and water. He says, "I approached the confines of death, and, having trod on the threshold of Proserpine, and I returned therefrom, being borne through all the elements."

As the aspirant thus wanders among these startling scenes, surrounded by the wild cries and lamentations of the goddess and her train, at a signal from the hierophant a sudden turn is given to their feelings. The gloom begins to disappear, and their cries of grief are changed to joyful and triumphant shouts of *"Eurekamen, eurekamen!"* ("We have found it!") The *euresis,* or discovery of the body, is then celebrated, and the mangled form of the murdered sun-god restored from death and darkness to life and light and power.

Another iron gate, heretofore concealed, is now thrown open. The Orphic hymn is chanted, and a splendid spectacle of the Elysian fields and the bliss of the purified presented. The four-and-twenty attendants of the hierophant prostrate themselves before him, and, amid strains of solemn music, the neophyte receives the benediction and instructions of the hierophant. (See Rev. A. C. Arnold's "History of Secret Societies"; Bishop Warburton on the "Mysteries"; Oliver's "History of Initiation"; Apuleius's "Metamorphoses"; and Salverti's "History of Magic")

The Mysteries of the *Cubiria,* or *Kabiri,* of Samothrace, were to the same effect, and were derived from the same Egyptian source—the Mysteries of Osiris and Isis—which they perhaps followed more closely. The candidate, after a term of probation, was purified by water and blood, made to sacrifice

16

**ISIS AND HORUS**

**GNOSTIC GEM OF ISIS**

"ISIS WAS WIFE OF OSIRIS AND
MOTHER OF HORUS. SHE WAS
ORIGINALLY THE GODDESS OF THE
EARTH, AND AFTERWARD OF THE
MOON. THE GREEKS IDENTIFY HER
BOTH WITH DEMETER, AND CERES,
AND WITH IO."
(SMITH'S "CLASSICAL DICTIONARY")

a bull and a ram, and to drink of two fountains, the one called
*Lethe* (oblivion) and the other *Mnemosyne* (memory), by
which means he lost the recollection of all his former crimes,
and preserved the memory of his new instructions and vows.
This is exactly similar to the Egyptian Mysteries. The candidate
was next conducted to a dark cavern, and thence through hor-
rible scenes similar to those before described. The walls were
clothed in black, and he was surrounded by all the emblems
of decay and death. Terrible phantoms passed and repassed
before him. A bier rose up at his feet, and on it was a coffin
and a dead body, representing the slain sun-god. A funeral
dirge was chanted by an invisible choir, and all the scenes of
terror multiplied.

**DIONYSUS
OR
BACCHUS**

**CERES,
DEMETER,
ISIS, ETC.**

These fearful visions were brought to a close by a flood dazzling light. All the emblems of death vanished. The dead body of the sun-god on the bier was raised and returned to life amid demonstrations of joy and triumph. The candidate was then instructed, sprinkled with water, and a new name given him. This new name, together with a mystic token and sign, was engraved upon a small white stone and presented to him.

## VIRGO

VIRGO IS THE SIGN THE SUN ENTERS IN AUGUST, AND WAS DEPICTED IN THE ZODIAC HOLDING IN HER HANDS THE EMBLEMS OF THE HARVEST. THE IDENTITY OF CERES, THE GODDESS OF THE HARVEST, WITH THE CONSTELLATION VIRGO, IS QUITE PLAIN. THIS FIGURE OF THE FRUITFUL VIRGIN WAS PLACED IN THE ZODIAC AS EMBLEMATIC OF THE HARVEST SEASON, BECAUSE THE SUN IS IN THOSE STARS AT THAT TIME. THE WORD "VIRGO" ORIGINALLY IMPLIED NOT ONLY A VIRGIN, BUT ANY VIRTUOUS MATRON. BY AN ASTRONOMICAL ALLEGORY THIS VIRGIN OF AUGUST BECAME A GODDESS, WHO DESCENDED TO THE EARTH, PRESIDED OVER THE HARVEST, TAUGHT MANKIND AGRICULTURE, AND WAS WORSHIPPED UNDER VARIOUS NAMES.

The Mysteries of Dionysus were the same as the Eleusinian and those of Bacchus, Dionysus being but one of the names of Bacchus.

The Dionysiac Mysteries and those of the Kabiri prevailed in Asia Minor, and spread through all the cities of Syria. *Hiram, King of Tyre,* was undoubtedly the high-priest of these Mysteries at Tyre, and the institution continued to exist in Judea as late as the time of Christ, as a secret society known as the *Essenes.* ("History of Secret Societies," by Rev. Augustus C. Arnold.

From the foregoing descriptions of the different Mysteries, it clearly appears that the main facts of the legend of the death of the sun-god and his return to life, as illustrated and celebrated in them all, are substantially the same, having been derived from the same source—the Mysteries of Osiris and Isis. The death of the sun-god, whom the "aspirant," dramatically represented, was the main characteristic of them all. So intimately were the ideas of death and initiation connected, that in the Greek language the same word expressed both ideas, τελευταν is *to die,* and τελεισθαι *to be initiated.*

(Warburton, "Div.Lg.," Book II, s. 4.) The *names,* however, by which the personified sun-god was known, varied with the language of the people:

> "Ogygia me Bacchum vocat;
> Osirin Egyptus putat;
> Mysi Phanacerp nominant;
> Dionuson Indi existimant;
> Romana sacra Liberum;
> Aribica gens Adoneum."
> —AUSONIUS, *Epigram* 30.

But, although the legend of initiation was thus substantially the same in all the civilized nations of antiquity, yet it must be borne in mind that the allegory of the death and return to life of the sun-god was naturally and necessarily modified in its minor details so as to conform to the different conditions of climate and order of the seasons, which prevailed in the various countries, into which it was adopted from Egypt. The Egyptians divided the year into seasons peculiar to themselves, consequent upon the exceptional nature of their country, where all agricultural pursuits were dependent upon and regulated by the yearly inundation of the Nile. They divided the year into three seasons of four months each: the first was called the season of *"Planets, "* and originally included November, December, January, and February; the second was termed the season of *"Flowering, " or "Harvest, "* and included March, April, May, and June; the third was known as the season of *"Waters, "* or *"Inundation, "* alluding to the overflow of the Nile, and originally consisted of July, August, September, and October. (Rawlinson's "Herodotus," vol. ii, page 238.) If we inscribe an *equilateral triangle* within the circle of the zodiac, placing *Taurus* on the vernal equinox, and *Leo* at the summer solstice, as was the case when the Egyptian seasons were first divided, we will have a correct representation of the ancient Egyptian year.

But, in the course of time, owing to the want of a correct knowledge of the true length of the solar year, these seasons

ANCIENT
EGYPTIAN
YEAR

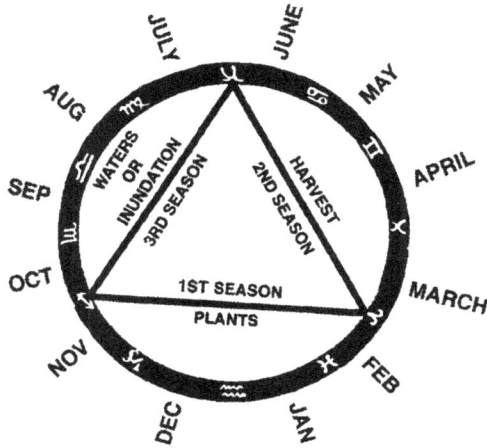

changed, and those of the summer fell in winter. It was therefore found to be necessary to make a correction of the calendar, which was done by observations taken of the heliacal rising of the dog-star *Sothis,* or *Sirius.* In their *sacred* calendar, however, the Egyptian priests appear to have retained the "vague" or indefinite year of three hundred and sixty days, so that the festivals of the gods illustrating the legend of Osiris might pass through all the different seasons of the year. (Wilkinson's "Ancient Egyptians," vol. ii, chapter viii.) This ignorance of the true length of the solar year produced also a similar confusion in the times of celebrating the festival of the gods in other countries, so that a festival, originally intended (for instance) to celebrate the arrival of the sun at the summer solstice, with appropriate ceremonies, might come to fall in winter, when the nature of those ceremonies had no harmony with the season. In like manner a festival, originally intended to celebrate the new birth of the sun at the winter solstice, would in process of time come to be held in the summer, and thus be in utter violation of the solar allegory. This, of course had the effect to entirely hide, or greatly obscure, the original solar allusion of these festivals, and it was probably for this reason that the Egyptian priesthood retained the "vague" year in their sacred calendar.

The neglect of the function of a year in the calendar does not appear to amount to much, but owing to this cause alone, the first of January in the time of Julius Caesar had fallen back so as to nearly coincide with the autumnal equinox. Caesar corrected the calendar, but, in order to do so, was obliged to make an extraordinary year of four hundred and forty-five days; this was called "the year of confusion." This correction made by Caesar did not prevent the recurrence of the same evil, for in process of time it was found that the seasons again began to disagree with the almanac, and the religious festivals of the Christian Church, like those of its pagan predecessor began to fall out of place. This led to the correction made by Pope Gregory, and the subsequent adoption of our present method of keeping the calendar correct. The solar allegory, when it was introduced into countries north of Egypt, and whose agriculture was not regulated by the overflow of the Nile, was modified, as we have seen, in some particulars, in order to harmonize the allegory with the climate and order of seasons which prevailed in those countries; but any want of correspondence that subsequently existed between the festivals, originally intended to celebrate the summer and winter solstices and the vernal and autumnal equinoxes, and the true time of the sun's arrival at those points, was due to an imperfect calendar, resulting from an ignorance of the length of the solar year.

Another cause which had the effect to obscure the original astronomical signification of the mythological tales of antiquity is the phenomenon known as the *"precession of the equinoxes, "* which has also changed the order of the seasons, so far as the same is marked by the entrance of the sun into particular constellations of the zodiac, at certain periods of the year. As, for instance, the advent of spring was anciently marked by the entrance of the sun among the stars of the constellation *Taurus,* it is now marked by his appearance among the stars of the constellation *Pisces.* The nature of this phenomenon and the astronomical changes which it has produced will be more fully explained in the following chapter.

In our astronomical explanation of the masonic traditions, legends, and emblems, all these causes thus tending to obscure and modify the original solar allegory, will be taken into account, and the same astronomically adapted, for obvious reasons to the astronomical conditions existing in countries north of the equator at the time of the building of King Solomon's temple, and some three or four hundred years immediately before and after.

Some of the masonic emblems, however, must be referred to a period much earlier, and some to a much later date, for it must be remembered that the astronomical legends and emblems of freemasonry did not all originate at the same period of time nor among the same people. They all, however, harmonize in their allegorical method, and strictly conform to the state of the heavens, and astronomical conditions, and the order of the seasons, as well as the degree of scientific knowledge of the era and country in which they respectively originated and become incorporated into that system of symbolical instruction then already existing, and now known as masonic.

It is the intention of this work to show— 1. That the masonic tradition is but one of the numerous ancient allegories of the yearly passage of the personified Sun among the twelve constellations of the zodiac—being founded on a system of astronomical symbols and emblems employed for the purpose of teaching and illustrating the two great truths, of the being of ONE spiritual, invisible, omnipresent, and omnipotent GOD, and the *immortality of the soul of man.* 2. That, while these two great doctrines were also originally taught in all the ancient Mysteries, by the use of the same astronomical allegories and symbols, freemasonry alone retained its primitive truth and purity, while the others degenerated into a corrupt system of solar worship. The sun, originally intended as a symbol only of the true God, was in time confounded with the person of God himself, and thus itself worshipped as a God. In freemasonry, on the contrary, it would appear that the exact reverse of this process has taken place, for, while the idea of God as an invisible spiritual being has been reverently kept alive, on the other hand, the original symbolism and primitive category relating to the sun as an illustration and emblem of the divine nature has been lost sight of, and the true meaning and profound scientific import of the masonic tradition, legends and emblems thus almost forgotten. The Rev. Dr. Oliver, whose great learning will be disputed by none, says:

> The poets, historians, and philosophers of Greece, *all of whom had been initiated* into the Mysteries, unite in describing the Supreme Being as ONE single, divine, and unapproachable essence, who created and governs the world. And in India the Supreme Deity is thus made to describe himself, in one of the sacred books, which has been preserved and transmitted from an unknown period: "I was even at first not any other thing; that which exists, the supreme; and afterward I am that which is; and he who must remain am I."
>
> ("Landmarks," Lecture XXI)

In the notes to this lecture of Dr. Oliver's, much valuable information on this point is also collected and condensed. The following is from the celebrated anthem of "Orpheus":

> When the doors are carefully guarded to exclude the profane, I will communicate the SECRET OF SECRETS to the aspirant perfectly initiated. Attend, therefore, to my words, for I shall reveal a solemn and unexpected truth to your startled ears—a truth which will overturn all your preconceived opinions, and convey to your mind unalloyed happiness. Let your soul be elevated to the contemplation of divinity. Adore Him, for He is the governor of the world. *Know that* HE IS ONE—that *He has no equal,* and that to Him all things are indebted for their existence. *He is everywhere present, though*

*invisible,* and all human thoughts are open to His inspection. (Note 27)

On the temple of Sais, in Lower Egypt, was inscribed the following sentence relating to the Deity:

> "I am all that hath been, and is, and shall be,
> And my veil no mortal hath yet removed."
>
> (Note 29)

In Note 32 to the same lecture, a translation is given of an extract from the Veda, which is deemed the oldest book in the world, except certain parts of the Bible. It is a translation made in 1656 by command of the Sultan Darah of an *Oupanishat,* a word meaning *the secret that is not to be revealed:*

> And what was this great mystery which was so carefully concealed in those ancient books? Like the secret of the *Egyptian and Grecian Mysteries,* it was nothing less than the *Unity of the Godhead,* under the name of *Ruder,* which is thus explained in another of their sacred books:

> The angels have assembled themselves together in heaven before *Ruder,* made obeisance and asked him, "O *Ruder,* what art thou?" *Ruder* replied: " *Were there any other,* I would describe myself by a similitude. I *always was, I always am, I always shall be.* There is no other, so that I can say to you, I am like him. In this ME is the inward essence and the exterior substance of all things. I am the *primitive cause* of all things in the east or west, or north or south—above or below, it is I. *I am* all. I *am* older than all. I *am* the King of kings. My attributes are transcendent. I *am Truth.* I am the spirit of creation. *I am the Creator. I am Almighty. I am Purity. I am the first, the middle, and the end. I am Light"*

Certainly no more sublime and comprehensive description of the eternal God was ever written.

Speaking of the antiquity of the Veda, Max Muller says:

> It will be difficult to settle whether the Veda is the "oldest of books," and whether some portions of the Old

> Testament may not be traced back to the same or even
> an earlier date than the oldest hymns of the Veda. But
> in the Aryan world the Veda is certainly the oldest
> book, and its preservation amounts almost to a marvel.
> (See "Lecture on the Vedas," at Leeds, 1865)

Muller in the same lecture fixes the date of the Vedas at "between twelve and fifteen hundred years before the Christian era." This is over three thousand years ago.

Dr. Oliver, in Note 34 to his lectures before quoted, informs us that Zoroaster taught that

> God is the first—incorruptible, eternal, unmade, invisible—most unlike everything—the leader or author of all good—unbribable—the best of the good—the wisest of the wise.

With all this evidence before him, and actually quoted in his writings, Dr. Oliver, strange as it may appear, is in the constant habit, in his works, of branding without distinction all the ancient Mysteries as "spurious freemasonry," an epithet which he invented, and which has been adopted by a few others. But, if the sublime views of God above quoted are "spurious," where shall we look for the genuine ones, for those taught in freemasonry today are the same?

Late discoveries make the fact, that the unity of God was taught in the ancient Egyptian Mysteries, beyond all doubt.

"The manifold forms of the Egyptian pantheon" (says the late E. Deutsch) "were but *religious masks* of the sublime doctrine of the unity of the Deity communicated to the *initiated* in the *Mysteries"* The gods of the Pantheon, says M. Pierrot, were "only manifestations of the One Being in various capacities." ("Dict. of d'Arch. Egypt.," article "Religion," Paris, 1875.) M. Maspero and other scholars arrived at the same conclusion. ("Hist. Anc. des Peuples de l'Orient," cap. i, Paris, 1876.)

The following hymn occurs in two papyri in the British Museum. It represents the thought prevalent in Egypt at the time of the Exodus, and is the work of Enna:

"Hail to thee, O Nile!

He causeth growth to fulfill all desires
He never wearies of it.
He maketh his might a buckler;
He is not graven in marble
As an image bearing the double crowns;
He is not behold;
He hath neither ministrants nor offerings;
He is not adored in sanctuaries,
His abode is not known.
No shrine is found with painted figures (of him).
There is no building that can contain him.
Unknown is his name in heaven.
He does not manifest his forms;
Vain are all representations of him."

And again we find the one God thus described:

He hath made the world with his hand, its waters, its atmosphere, its vegetation, all the flocks and birds, and fish, and reptiles, and beasts of the field.
(Hymn to Osiris, translated by Chabas)

"He made all the world contains, and hath given it light when there was yet no sun." (Melange's "Egypt," i, 118, 119. Chabas.)

Glory to thee who hast begotten all that exists, who hast made man, and made the gods also, and all the beasts of the field! Thou makest men to live. Thou hast no second to thee. Thou givest the breath of life. Thou art the light of the world.
(Leeman, "Monuments du Musee des Pays-Bas," ii, 3)

But although God was the creator, yet he is himself "self-created."

His commencement is from the beginning. He is the God who has existed from old time, There is no God without him. No mother bore him, no father hath begotten him. God-goddess created from himself.
(Chabas)

In many of the hymns we find allusion made to the mystery of his name, and its being hidden, secret, and unknown—ineffable, and not to be spoken.

"Unknown is his name in heaven. Whose name is hidden from his creatures. His name which is Amen" (i.e., hidden secret). Therefore the Egyptians *never spoke the unknown name,* but used a phrase which expressed the self-existence of the eternal, "I am," (Ritual of the Dead.)

Says John Newenham Hoare, in the late article in the "Nineteenth Century":

> The Egyptians tried to realize God by taking some natural object which should in itself convey to their minds some feature in God's nature. This became a necessity for the priests in the religious teaching of the people. Therefore in the sun they saw God manifested as the light of the world. The more fully they felt the infinite nature of God, the more they would seek in nature for symbols.... All the deities were regarded as manifestations of the one great Creator, the uncreated, the Father of the universe.

This is expressed in the following hymn:

> Hail to the Lord of the lapse of time, king of gods! Thou of *many names,* of holy transformations, of mysterious forms.

> Nevertheless, as in Greece and India, so also in ancient Egypt, the *symbols became* in the popular mind actual gods, and the people degenerated into gross idolatry.

> "They *changed* the glory of the incorruptible God into an image made by corruptible man, and to birds, and to four-footed beasts, and creeping things,... and they *changed* the truth of God into a lie, and worshipped and served the creature rather than the creator."
>
> (Rom. 1:23-25)

> This is unfortunately the aspect in which the Egyptian Pantheon has presented itself to mankind for many centuries.

The conception of the unity of the Godhead did not prevent the Egyptians from thinking of God as very near to them.

He is their father, and they "sons beloved of their father." He is the "giver of life," "toucher of the hearts," "searcher of the inward parts is his name."

Every one glorifies thy goodness; mild is thy love toward us; thy tenderness surrounds our hearts; great *is* thy love in all the souls of men.

One lamentation cries:

Let not thy face be turned away from us; the joy of our hearts is to contemplate thee. Chase all anguish from our hearts. He wipes tears from off all faces. Hail to thee, Ra! Lord of all truth, whose shrine is hidden.

Lord of the gods, who listeneth to the poor in his distress, gentle of heart when we cry to thee. Deliverer of the timid man from the violent, judging the poor—the poor and oppressed. Lord of mercy, most loving; at whose coming men live, at whose goodness gods and men rejoice—sovereign of life, health, and strength.

("Records of the Past," ii, 98)

Speak nothing offensive of the Great Creator; if the words are spoken in secret, the heart of man is no secret to him that made it.               (Ibid., ii, 131)

He is present with thee though thou be alone.

As we might expect, from so lofty a conception of God, their hearts broke forth into joyous hymns of praise:

"Hail to thee, all creatures!
Salutation from every land.
To the height of heaven, to the breadth of the earth,
To the depths of the sea,
The gods adore thy majesty.
The spirits thou hast made exalt thee,
Father of the father of all the gods,
Who raises the heavens, who fixes the earth.
Maker of beings, author of existences,

Sovereign of life, health, and strength,
           Chief of the gods,
We worship thy spirit, *who alone* has made us.
We, whom thou hast made, thank thee that thou hast
           given us birth.
We give thee praises for thy
           Mercy toward us."
                        ("Records of the Past," ii, 98)

Such was the idea of God and his relations to man held by the ancient Egyptians, and, as we might expect, it drew forth in them "lovely and pleasant lives." The three cardinal requirements of Egyptian piety were— love to God, love of virtue, and love to man.

The honor due to parents sprang naturally from the belief in God as "Our Father, which art in heaven." We constantly find inscriptions on the tombs such as the following: "I honored my father and mother; I loved my brothers; I taught little children; I took care of orphans as though they had been my own children." In letters of excellent advice, addressed by an old man one hundred and ten years of age to a young friend (which forms the most ancient book in the world, dating 3000 B.C.), he says: "The obedience of a docile son is a blessing. God loves obedience. Disobedience is hated by God. The obedience of a son maketh glad the heart of his father. a son teachable in God's service will be happy in consequence of his obedience. He will grow to be old, he will find favor."

That our ancient brothers of Egypt were not deficient in the masonic virtues of "BROTHERLY LOVE AND RELIEF AND TRUTH," appears from the following:

On the tombs we find the *common formula:*
"I have given bread to the hungry, water to the
thirsty, clothes to the naked, shelter to the stranger."
This tenderness for suffering humanity is characteristic
of the nation.

Gratefully does a man acknowledge in his autobiography (4000 B.C.) that, "Wandering, I wandered and was hungry; bread was set before me; I fled from the land naked; there was given me fine linen."

(Chabas, "Les papyras Hieratiques de Berlin; revits d'il y a quatre mille Ans," 1863)

Love of truth and justice was also a distinguishing trait of the Egyptians. God is thus invoked, "Rock of truth is thy name." In an inscription at Sistrum a king addresses Hathor, goddess of truth, "I offer to thee the truth, O goddess, for truth is thy work, and thou thyself art the truth." Truthfulness was an essential part of the Egyptian moral code, and in the Egyptian Ritual we are informed that, when after death the soul enters the hall of the Two Truths, or Perfect Justice, it repeats the words learned upon earth: "O thou great God, Lord of Truth, I have known thee, I have known thy name: Lord of Truth is thy name. I never told a lie at the tribunal of truth."

("Religion of the Ancient Egyptians," by John Newenham Hoare, a late article in the "Nineteenth Century")

But enough had been advanced to establish the fact that the ancient Mysteries originally taught the unity of God, and also that their moral code was both pure and exalted.

That the ancient Mysteries, after the people became corrupt, became corrupt in their turn, there can be no doubt, but in their inception they were not so. The crowning secret was a knowledge of the true God, and the disclosure of the fact that the sun was only a symbol of the great Creator, and not itself a divine being. In the midst of an age where the worship of the sun was the established religion of all nations, no one could with safety avow his disbelief in the divine nature of the heavenly bodies. To do so would be instant destruction.

Before the great truth of the real nature and attributes of God could be communicated, the candidate was required to take all the degrees of the Mysteries, and give the strongest proofs of his fidelity and zeal.

A knowledge of the true God was, in the language of the Orphic hymn, "the *secret of secrets*" to be only communicated when the aspirant was *"perfectly* initiated," with *"doors care-fully guarded* and the *profane excluded!"* It was even then, to those to whom it was to be communicated, *"a solemn* and *unexpected* truth which "startled their ears" and *"overturned their preconceived opinions."*

Taught from their earliest infancy to regard the sun, moon, and stars as actual divinities, a wandering in the darkness of a false system religion, they were on their initiation into the Mysteries first brought to behold the true light and there obtained for the first time a knowledge of the true God. This was the real AUTOPSY, "bringing to light," of the candidate in the Mysteries. "It was difficult," says Plato, "to attest and dangerous to publish, the knowledge of the true God.

The light thus communicated under the strictest conditions of secrecy to be kept, when communicated, religiously hidden from the initiated, it being well known that a public profession of the great truth would be visited by a heavy hand of both the civil and religious authorities, and not only their own lives but that of their kindred be thus sacrificed to the superstitious rage of the ignorant multitude, and the interested fury of the ministers of a false religion.

It is true that the priests themselves often took an active part in the Mysteries, of which they had taken the higher degrees. The Mysteries served as a sort of theological and scientific seminary, which they studied the truths of religion and science, and from the higher degrees of which the ranks of the priesthood and rulers were from time to time recruited. But these facts could be of a help to him who rashly made a *public* profession of his want of faith in the national solar gods.

The policy of secrecy, by which all but truth, whether religious or scientific, was concentrated in and confined to the Mysteries, was a "stated policy" long established and though to be necessary for the well-being of society. It certainly was for the well-being of the few on whom it conferred power and wealth. To "reveal the Mysteries" was considered the very highest of crimes and he who did so could hope for no mercy. The very priests who perhaps had initiated him, and who did not themselves believe in the divinity of the sun, moon, and stars, would be the first to denounce his alleged impiety and atheism, and urge on his punishment. Nor would any of the brotherhood help him, as he would be considered by them as a perjured traitor, who had violated the most solemn obligations, and now sought to destroy the order itself by exposing it to the superstitious wrath of the ignorant multitude.

> The betrayers of the Mysteries were punished capitally and with merciless severity. Diagoras the Melian had revealed the Orphic and Eleusinian Mysteries, on which account he passed with the people as an atheist, and the city of Athens proscribed him and set a price on his head. The poet AEschylus had like to have been torn in pieces by the people, on the mere suspicion that in one of his scenes he had given a hint of something in the Mysteries. (Warburton)

So long, however, as the initiated held their peace, they all might, at the solemn assemblies of the Mysteries, held under circumstances of profound secrecy and sanctioned by the government itself, worship the one true God without fear; indeed, such a worship was enjoined upon them. But, should they openly disclose their belief in the actual divinity of the sun, moon, and stars, their danger was immediate and their ruin certain. Thus all alike, from the most exalted hierophant to the humblest of the initiated, were the slaves, and sometimes the victims, of a system of state policy which they all upheld and defended. It is true, however, that in the progress of many centuries the Mysteries became corrupt, and lost a knowledge of the true God, but in their original institution they not only taught the truth concerning the Deity, but protected his worshipers so long as they kept sacred their vows of secrecy. That the doctrine of immortality was also directly taught in the Mysteries, we are informed by Cicero, who had himself been initiated. (See "Tusculan Disputations," Book I, cxiii.) Among all the corruptions which at a later date prevailed, there, however, yet remained a "chosen band," who preserved the ancient teachings of the Mysteries in their purity. They were obliged for their own protection, however, to render their symbols yet more obscure, and make thicker and draw still closer the veil of allegory about the *penetralia* of divine truth. From these few and faithful ones the truth was handed down to following generations, and from them all that is great, glorious, and ancient in modern freemasonry was derived.

From those freemasonry received its two great doctrines of the unity of God and the immortality of man; and, together with those sublime truths, it also received that system of astronomical symbols, emblems, and allegories also peculiar to the Mysteries, which were used, anciently, both to conceal and to illustrate those great truths. Dr. Mackey, in his "Symbolism of Freemasonry," says that those to seek for an astronomical explanation of the masonic ritual, "yield all that masonry gained of religious development in past ages" (page 237). For this broad assertion he gives no reasons whatever, and I can not but think that, had he considered the full import of his words, he never would have made any such remark. There is certainly nothing irreligious or atheistical in the employment of astronomical emblems to describe and illustrate the nature and attributes of Deity. If so, the writers of the Bible have been guilty of a great sin, for that sacred volume is full of solar and astronomical illustrations of the glory and power of the creator. (Numb. 24:17; Psalm 19; 84:11; Mal. 4:2; Matt. 2:2; 17:2; Judges 5:20; Job 25:5; 38:7; Dan 12:3; Jude 13; Rev. 1:16; 10:1, etc.) Freemasonry, says Dr. Mackey, quoting Dr. Hemming with approval, is a science of morality "veiled in allegory and illustrated by symbols." It is to be inferred that the moral science taught in freemasonry is any the less true, pure, or elevated, because the allegories and symbols employed to "veil and illustrate" it are astronomical in their character? Is it irreligious and atheistical to compare the great Creator to the
34

noblest and most glorious of all his physical works—the sun—and only orthodox and pious to compare his nature and attributes to a carpenter's rule or a stonecutters square? Certainly this is not what Dr. Mackey intends, yet such is the natural inference from his language.

Neither does it follow that those who give the masonic ritual an astronomical and scientific as well as a moral interpretation, deny to masonry the glorious distinction of having been in past ages the depository of a knowledge of the true God, and of the immortal nature of man. All that we contend is, that those great truths were taught not only by allegory and symbol, but originally and mainly by *astronomical* symbol and allegory.

The more exalted and holy any doctrine is, the more elevated and sublime should be the symbols and emblems to teach and illustrate it.

As the being and attributes of God and the immortality of the soul are the two most exalted and sublime of all truths, so are the sun, moon and stars the most glorious and sublime objects in nature. There are, therefore, a peculiar fitness and beauty in the employment of the latter to symbolically and emblematically illustrate the former. "The heavens declare the glory of God, and the firmament showeth his handiwork."

In this work no attempt will be made to identify the masonic emblems, traditions, and legends with the Mysteries of any particular nation. All the Mysteries were originally astronomical in their character, but differed in form and detail, as they were founded on different modifications of the Egyptian legend of the personified sun-god. Dr. Mackey, in strange contradiction to the words which we have above quoted from page 236 of his "Symbolism of Freemasonry," devotes a whole chapter of that interesting and learned work to prove that freemasonry was derived directly from the Grecian Mysteries of *Dionysus.* He thinks it certain that the Tyrian artificer, Hiram, was a member of the Dionysiac fraternity, and that he, at the head of the Tyrian workmen at the time of the building of King Solomon's temple, introduced the Dionysiac Mysteries in a modified and purified form among the Hebrews (Chapter VI). Dr. Oliver, who denies in all its detail the astronomical theory, with an equal inconsistency advocates the same idea. (See his "Theocratic Philosophy of Freemasonry," Lecture VIII.) According to Dr. Mackey, and Dr. Oliver, also, freemasonry is therefore only a modified and purified form of the Grecian Mysteries of Dionysus.

It is true that, like the others, these Mysteries became corrupt, but is equally true that the Mysteries of Dionysus, like all the other Mysteries, were astronomical in their character. Dionysus is but another name for Osiris, and is the personified sun-god, the legend of whose death, the search for whose body, and its recovery, together with his subsequent "raising" from death and the grave of a new life, forms the theme of the ceremony of initiation; all of which the aspirant was caused to dramatically enact.

"One thing, at least" says Dr. Mackey,

> is *incapable of refutation;* and that is, that we are indebted to the Tyrian masons for the introduction of the *symbol* of Hiram Abif. The idea of the *symbol,* though modified by the Jewish masons, is not Jewish in its inception. It was evidently borrowed from the pagan "Mysteries," where Bacchus, Adonis, Proserpine, and a host of other apotheosized beings *play the same role* that *Hiram does in the masonic Mysteries.*
>
> ("Symbolism of Freemasonry," Chapter I, page 20)

This emphatic language of Dr. Mackey, therefore, not only admits, but declares *"incapable of refutation,* "the following important particulars:

1. That Hiram Abif, as described in the masonic legend, is a mystical being, or "symbol" only, and not a historical person, any more than Bacchus, Adonis, or Proserpine.

2. That the whole legend of the third degree is an allegory and not a history.

3. That the allegory is the same as that of Bacchus, or Dionysus, and therefore identical with that of Osiris. (For proof that the Mysteries of Bacchus, or Dionysus, were the same as those of Osiris, see "Herodotus," Book II, Chapter LI, sections 49-60; together with the notes to Rawlinson's edition. Also, as to the identity of Bacchus and Dionysus, see Oliver's "History of Initiation," Lecture VI, and notes.)

4. That in this allegory Hiram "plays the same *role* as that of Bacchus, or Dionysus, and Osiris, and all the other personified sun-gods in the various forms of the Mysteries.

Now what is this *role?* It is simply that of the personified sun—slain like Osiris, Bacchus, Adonis, or Dionysus, at the Autumnal equinox; lying dead during the winter months, being restored to life at the vernal equinox, and exalted in power and glory at the summer solstice.

These admissions of Dr. Mackey cover the whole ground, and sanction every position to be taken in this work. It is not, however, my intention to trace the masonic traditions, legends, and emblems, like Dr. Mackey, to any one of the ancient Mysteries to the exclusion of the others, as masonry has features derived from each of them. It *is,* however, my design to show that it is of an *astronomical* nature, and had its origin, in common with all the ancient Mysteries, in a lofty system of *astronomical allegories,* originally intended to teach the unity of God, the immortality of the soul, and an exalted code of morality; while at the same time, by the use of the same allegories and symbols, the leading facts of astronomical science were to be both illustrated and preserved—in other words, to show that *freemasonry is a system of science as well as morality, veiled in an astronomical allegory, and illustrated by astronomical symbols.*

It is also the intention of this work to unlock this allegory, and to show the true scientific and astronomical meaning, as well as moral application, not only of all the legends, but of all the emblems and symbols of freemasonry which have any claim to antiquity.

The real character and true origin of the peculiar symbolism of freemasonry and its allegories have been a great puzzle to most members of the fraternity. The great moral truths which those symbols and allegories teach are plain enough; the only mystery is, how came those truths to be taught by those peculiar symbols and in that peculiar manner?

It is also worthy of remark that, while the moral truths which our emblems, symbols, and legends teach are still well understood, yet those great scientific truths, which are equally said to illustrate and teach, are wholly lost, and at least their connection with them. This lost connection between our emblems, symbols, and legends, and many of the profoundest truths of science, will be restored in the pages of this work.

Oliver and Hutchinson have both, with much labor, and the former with great learning, attempted to prove that the master-mason's degree is a Christian institution—not in the sense of its being pervaded with the spirit of Christianity, which is true, but a Christian institution in the same sense as the Church or the rite of baptism is. Dr. Mackey correctly says they have "fallen into a great error." The theory that our fraternity had its origin in the building societies of the middle ages is sufficiently disproved by our ritual itself, which has many features that are totally inconsistent with any such theory, and point to a far more remote era; although many things relating to operative masonry were no doubt then ingrafted on it.

Dr. Mackey, Oliver, and others, will not accept the astronomical theory, and thus the whole matter remains, so far as they are concerned, a mystery. The astronomical theory is, however, the only correct one, as the following pages will sufficiently show.

The great difficulty is, that it has never been properly and at the same time fully presented. It has been advanced mainly by antimasons, who understood many other things much better than they did our ritual and the legends and symbolism of our order; or by skeptics, endeavoring at the same time to tear down the Christian religion. The advocacy of the astronomical theory by this kind of writers, especially the latter, has done much to render it unpopular, and induced many authors and thinkers to discard it without a due and fair examination. Many masons, like Dr. Oliver, seem to have an illogical and almost superstitious fear of having the astronomical character of our symbolism established. The fact is, however, that the great moral truths of freemasonry are indestructible, and stand independent of the symbolism intended to illustrate them, and to conceal them also, in past ages, when disclosure exposed the initiated to persecution and death, as an unbeliever in the actual divinity of the sun, moon and stars. The great moral teachings of freemasonry will not suffer any danger of destruction or damage if it *is* fully established that the emblems by which they are illustrated, like the imagery of the Bible, are mainly astronomical instead of mechanical.

The following pages, it is believed, contain convincing proofs of the real character and origin of our symbolism. Portions of the masonic ritual, and a few of the emblems, have in a general way been shown by several writers to be of astronomical origin, and the assertion has been frequently made that the whole system has an astronomical significance. But it is believed that this work contains the only full and complete demonstration of the purely astronomical and scientific import of the *whole* ritual, and all the details of the solar allegory, as applied to masonry—accompanied by a particular exposition of the astronomical import and origin of *all of its ancient emblems, symbols, and legends,* over seventy in number (see index), that has ever been made. The traditions and emblems of freemasonry have been made to speak for themselves, and they tell their own origin and meaning in a language which can not fail to convince any reader, who combines a knowledge of the lodge and chapter degrees with the main outlines and leading principles of astronomy and geometry. These sciences, so often alluded to in our ritual, are eminently masonic, and without some knowledge of them what is to follow will not be fully understood.

It is hoped that this work will also not be without interest to the uninitiated. They will at least, be able to see, unfolded in its pages, a beautiful and impressive astronomical allegory, which, by the use of sublime and august emblems, teaches the unity of God and the immortality of the soul. The work also throws much light upon the religion of the ancient Egyptians, Greeks and Romans, as well as mythology in general. How far the solar allegory may be truthfully applied to freemasonry they, of course, will not be able fully to determine for themselves, except in a general way and on minor points. As for the rest, they will be expected to be complacent enough to take the opinion of well-informed members of the fraternity.

# Chapter 3

# *A Chapter of Astronomical Facts*

## IN ORDER TO PROPERLY UNDERSTAND

what is to follow, some knowledge of the leading facts of astronomy is required. The nature of the *zodiac,* and its division into signs and constellations; the phenomena attending the yearly passage of the sun among the stars; the solstitial and equinoctial points, and the "precession of the equinoxes," and its effect upon the relative position of the signs and constellations of the zodiac—as well as several other particulars of astronomy—must be known by the reader, in order that he may fully understand the astronomical allegory about to be unfolded and illustrated.

It has, therefore, been thought necessary to write an introductory chapter, giving a brief and popular exposition of the matters above enumerated. All technical terms will be discarded, as far as possible, and such as are used from absolute necessity will be defined. No attempt will be made to give a cause or philosophy of solar or sidereal movements—the sole object being to bring clearly before the mind the *apparent* annual path of the sun in the zodiac, and such other celestial phenomena as are required to properly understand the allegorical application which is to be made of the facts of astronomy to the masonic traditions, legends, emblems, and symbols. This chapter will serve to call the particular attention of those who are proficient in science to certain particular astronomical facts bearing directly upon our subject, and it is hoped will also contain enough to sufficiently instruct those who may have grown rusty in or never acquired a knowledge of the motions of the celestial bodies.

### The Ecliptic

The ecliptic is a great circle in the heavens surrounding the earth, and representing the apparent path of the sun each year among the stars.

### The Zodiac

The zodiac is a belt of stars extending 8° on each side of the imaginary circle called the ecliptic. The zodiac is therefore 16° wide, and, being a complete circle, is 360° in circumference. It is divided into twelve equal parts of 30°, each denoting the particular place which the sun occupies during each of the twelve months of the year. Each of these divisions of the zodiac, in the visible heavens, is marked and occupied by a separate and distinct group or cluster of stars, called a constellation. These constellations are named after certain "living creatures," supposed to have been originally emblematic of the month in which the sun entered them.

The twelve constellations are called—

**TABLE 1.**

| | |
|---|---|
| *Aries,* the Ram. | *Libra,* the Scales. |
| *Taurus,* the Bull. | *Scorpio,* the Scorpion. |
| *Gemini,* the Twins. | *Capricornus,* the Goat. |
| *Cancer,* the Crab | *Sagittarius,* the Archer. |
| *Leo,* the Lion | *Aquarius,* the Water-Bearer. |
| *Virgo,* the Virgin | *Pisces,* the Fishes. |

These, ranged in their appropriate places in the great zodiacal circle, are all represented in the foregoing diagram. The following is a brief description of each of the constellations:

## Aries

This was once the first constellation of the zodiac. It is now the second, by reason of the precession of the equinoxes, which will be subsequently explained. It is known by two bright stars, about 4° apart, which are in the horns of the Ram. The brightest of these, called *Arietus,* is used by navigators to compute longitude by the moon's distance. Most of the stars in this constellation are small. Aries, in the Hebrew zodiac, is assigned to Simeon, or by some to Gad.

## Taurus

This constellation is next to *Aries* in the zodiac, and is one of the most celebrated and splendid. The *Pleiades* are in *Taurus,* and near it is the magnificent constellation *Orion,* called *Orus* by the Egyptians. In that sublime chapter of the Old Testament, Job xxxviii, mention is made of these: "Canst thou bind the sweet influences of the *Pleiades,* or loose the bands of *Orion?" Taurus,* once seen and recognized in connection with *Orion,* is never forgotten.

The Bull is represented as engaged in combat with *Orion,* and plunging toward him with threatening horns. The face of the Bull is designated by five bright stars in the shape of a letter V, known as the *Hyades,* the most brilliant of which is *Aldebaran,* which is much used by navigators. The tips of the horns of the Bull are marked by two bright stars at an appropriate distance above the face. The *Pleiades* gleam brightly near the shoulder. *Orion,* who faces the Bull, is known by four bright stars, forming a large parallelogram, in the center of which is seen a diagonal row of stars, known as the belt of *Orion,* and called in Job the "bands of Orion." The four stars of the parallelogram, respectively, indicate his shoulders and feet. A line of smaller stars form his sword, its handle ornamented by a wonderful *nebula. Just* below *Orion* shines, with a splendor almost equal to Jupiter or Venus, that mighty sun-star *Sirius,* the deified *Sothis* of the Egyptians. Further east and over him flashes that brilliant star known as *Procyon.* These two, with *Betelgeux,* in the shoulder of *Orion,* form an equilateral triangle, whose sides are each 26°, which is so perfect and beautiful as almost to force itself upon our attention. *Taurus, Orion, Sirius,* the *Pleiades,* and *Hyades,* are all frequently alluded to by the poet Virgil in the "Georgics." This is, perhaps, the most magnificent and sublime quarter of the heavens north of the equator.

*Taurus* was held by the Egyptians, and most of the nations of antiquity, as a sacred constellation. Before the time of Abraham, or over four thousand years ago, it adorned and marked the vernal equinox, and "for the space of two thousand years the Bull was the prince and leader of the celestial host." The sun in Taurus was deified under the symbol of the bull, and worshipped in that form. The sacred figures found among the ruins of Egypt and Assyria, in the form of a bull with a human face, or with a human shape with the face and horns of a bull, are emblematic of the sun in Taurus, at the vernal equinox. In the Hebrew zodiac Taurus was ascribed to Joseph.

### Gemini

—is the next constellation in the zodiac. Its principal stars are two bright ones, called *Castor* and *Pollux.* They are about $4^{1}/_{2}°$ apart, and of the first and second magnitudes. In mythology, Castor and Pollux are said to be twin sons of Jupiter by Leda. In the Hebrew zodiac this constellation is assigned to Benjamin.

### Cancer

This constellation is composed of two stars, the brightest of which are only of the third magnitude. It is of no especial importance, except from its position, of which more will be said subsequently. In some Eastern zodiacs this sign is

44

represented by the figure of two animals like asses, and by the Hebrews is assigned to *Issachar.*

## *Leo*

This is another celebrated and beautiful constellation. It is easily known by five or six bright stars situated in the neck and head of the Lion, and arranged in the form of a sickle. Its two brightest stars are *Regulus* and *Denebola,* the former in the sickle and the latter near the tip of the tail. *Regulus* is a very bright star, and is situated almost exactly in the ecliptic. It is, therefore, of great use to navigators in determining the longitude of the sea. The constellation *Leo* is also celebrated as being the radial point from which the remarkable meteoric showers of November proceed. If this phenomenon was observed by the ancients, it must have greatly increased the veneration and awe with which this sacred constellation was viewed.

The constellation Leo is, for many reasons, full of significance to masons. It once marked the summer solstice, and at the building of King Solomon's temple was much nearer that point than now; this change of position, consequent upon the precession of the equinoxes, will be subsequently explained, together with the intimate connection between the constellation Leo and the masonic tradition. In the Hebrew zodiac Leo is the significator of the tribe of Judah. According to astrology, it is the "sole house of the sun."

## *Virgo*

This is the beautiful virgin of the zodiac. She is represented as holding a spear of ripe wheat in her left hand, marked by a brilliant star, called *Spica.* In the Egyptian zodiac Isis supplies the place of Virgo, and is represented holding three ears of corn in her hand. Spica, together with *Denebola* in Leo, and *Arcturus* in Bootes, forms an equilateral triangle of great beauty. *Arcturus* is also one of the stars mentioned by Job: "Canst thou bring forth *Mazzaroth* in his season? or canst thou guide *Arcturus* with his sons?" *Mazzaroth* signifies the twelve signs of the Zodiac. *Arcturus* is also frequently alluded to by Virgil in the first book of the "Georgics." The rising and setting of this star were supposed to portend great tempests. In the time of Virgil it rose about the middle of September. The bright star *Spica,* in Virgo, lies within the path of the moon, and is of great use to navigators. In the Hebrew zodiac Virgo is assigned to *Naphtali,* whose standard was a tree bearing goodly branches.

## Libra

This constellation is anciently represented by the figure of a man or woman holding a pair of scales. The human figure is omitted in all Arabian zodiacs, as it is held unlawful by the believers in the Koran to make any representation of the human form. In our zodiac, also, the balance only is depicted, probably because we received the zodiac from the Arabians. This constellation may be distinguished by a quadrilateral of four stars, but it contains none of great brilliancy. In the Hebrew zodiac Libra is ascribed to *Asher* This constellation formerly was on the autumnal equinox, and when the sun entered its stars the days and nights were equal. To this the Latin poet Virgil alludes:

> Libra die somnique pares ubi fecerit horas,
> Et medium luci atque umbris jam devidit orbem.

> When Libra makes the hours of day and night equal, and now divides the globe in the middle, between light and shades.　　　　　— "Georgics," Book I

## Scorpio

this constellation has some resemblance, in the grouping of its stars, to the object after which it is named. It is a very conspicuous object in the evening sky of July. In its general form it resembles a boy's bow kite, the tail of which forms that of a scorpion, and is composed of ten bright stars. The first of these, near the point of the triangle forming the body of the kite, is *Antares.* It is a brilliant red star, resembling the planet Mars. In the Hebrew zodiac Scorpio is referred to Dan.

### Sagittarius

The Archer follows Scorpio, and is represented as a monster, half horse and half man, in the act of shooting an arrow from a bow. *Sagittarius* is easily recognized by the figure of an inverted dipper, formed of several bright stars. The figure of Sagittarius appears in the ancient zodiacs of Egypt and India.

### Capricornus

The Goat is composed of fifty-one visible stars, most of them small. It is of no particular importance, except from the connection of its sign with the winter solstice, of which more will be said hereafter. It was called by the ancient Oriental nations the *southern gate* of the sun.

## Aquarius and Pisces

These are the last two constellations of the zodiac. The former is represented by the figure of a man, pouring out water from a jar, the latter by two fishes joined at a considerable distance by a loose cord. *Aquarius* in the Hebrew zodiac represents the tribe of Reuben, and the Fishes Simeon. The stars in both of these constellations are small and unimportant, except *Fomalhaut,* in *Aquarius,* which is almost of the first magnitude, and is used by navigators. This concludes our description of the *constellations* of the zodiac.

### The Signs of the Zodiac

The *signs* of the zodiac are twelve arbitrary signs, or characters, by which the twelve constellations are designated. They are as follows:

♈ ♉ ♊ ♋ ♌ ♍ ♎ ♏ ♐ ♑ ♒ ♓

These, without doubt, had their origin in the hieroglyphic or picture writing of the ancients. In the sign Aries (♈) we have a rude but yet remaining representation of the head and horns of the Ram. In Taurus (♉) of the face and horns of a

Bull. Gemini (♊) denotes the Twins, seated side by side with embracing arms. The ancient statues of Castor and Pollux consisted of two upright pieces of wood, joined together by two cross-pieces. Cancer (♋) yet retains a resemblance to the claws of the Crab. Leo (♌) may be intended for a crouching lion, or may be the outline of its principal stars—the group now called the Sickle, the stars of which, if joined by an imaginary line, would form a figure not unlike the sign (♌). In Virgo (♍) the resemblance seems to be lost. Libra (♎) is a plain picture of a scale-beam. The sign Scorpio (♏) displays the sting of a venomous creature. Sagittarius, the Archer, is well represented by his arrow and part of his bow (♐). In Capricornus (♑) the resemblance is again lost; but in Aquarius (♒) we recognize the waves of the sea, denoting water. In Pisces (♓) the resemblance of two fishes joined is still apparent.

It is quite easy to conceive how the original pictorial representations of the creatures emblematically denoting the various constellations of speed and convenience in writing them, grew into these arbitrary signs like letters.

In the figure of the zodiac, opposite page 42 the pictorial representations of the twelve constellations are given, with the arbitrary signs denoting each placed against them. The sun, moon, and planets were also designated by hieroglyphic astronomical signs by the ancients, as follows.

| | | | |
|---|---|---|---|
| *Sun,* | ☉. | *Venus,* | ♀. |
| *Moon,* | ☽. | *Jupiter,* | ♃. |
| *Mercury,* | ☿. | *Saturn* | ♄. |
| *Mars,* | ♂. | | |

The planetary signs originated in the same manner as the zodiacal ones. The sign for the sun is "a point within a circle"—the point represents the earth, and the circle the ecliptic. The moon is appropriately pictured as a crescent. In the sign of Mercury we have the *caduceus* of that god, composed of two serpents twisted about a rod. Mars is represented by his shield and spear. Venus is well denoted by the picture of an ancient hand-mirror. The origin of the planetary sign for Jupiter is not so clear. It does not in the least resemble an eagle, as some suggest, nor is it any more like the initial letter of the Greek *Zeus;* besides, the hieroglyphs are always representations of objects, not letters. This sign resembles more nearly the no less ancient numeral sign, the figure 4, and, as Jupiter is the fourth planet from the sun (if, like the ancients, we do not enumerate the earth), this resemblance may not be accidental. Saturn, lastly, is represented by his scythe in its ancient form.

These arbitrary signs for the planets and constellations have come down to us from a remote antiquity. Their general use, by all civilized nations is of great benefit, as they form a kind of astronomical shorthand, which, like the Arabian or Hindoo numerals, is equally well understood in all countries, no matter how much their language or ordinary written characters may differ, so that astronomical tables for the use of navigators and others are as well understood and as easily read in any one part of the civilized world as another. The great convenience of this is so apparent as to require no comment. The time when the zodiac was divided into twelve constellations, and the zodiacal signs invented, is lost in the dim distance of an extreme antiquity. The best opinion at present seems to be that the zodiac was derived from the Hindoos by the Egyptians, who gave it to the Arabians, who preserved it, and in turn transmitted it to us. Baldwin, in his "Prehistoric Nations," however proves that it is highly probable that the ancient Arabians originated it in prehistoric times. When the signs of the zodiac first began to be used, or what ancient students of the starry skies invented them, is therefore unknown, save by conjecture.

The zodiac has four principal points: these are the two *solstitial and* two *equinoctial* points, which, dividing the circle of the zodiac into four equal parts, are properly designated in the foregoing diagram. These four points were anciently marked by the stars *Fomalhaut, Aldebaran, Regulus,* and *Antares.*

## The Solstitial Points

The solstitial points refer to the movement of the sun, north of the equator in summer and south of it in winter. They are the points marking the extreme northern and southern limits of this movement of the sun. The *summer solstice,* when the sun comes farthest north, is at present in *Cancer,* and the *winter solstice,* or his extreme southern limit, is in *Capricornus.* The distance of the sun north and south of the equator is called his northern or southern *declination.* When the sun reaches either solstitial point, he begins to turn back toward the other—at first very slowly, and for a short period seems to stand still. It is for this reason that these points are called "solstitial," from the latin words *sol,* the sun, and *sistere-stiti,* to cause to stand. When, in June, the sun enters *Cancer,* and reaches his greatest northern declination, his rays, falling more vertically, cause the change from winter and spring to summer in all countries north of the equator. This shifting of the sun from one solstitial point to the other is the cause of the change of the seasons.

## The Equinoctial Points

These are the points where the sun crosses the celestial equator, which he necessarily does twice in his yearly circuit of the zodiac, at two opposite points, distant from each other in space 180°, and in time six months. The point where the sun crosses in spring, coming north, is called the *vernal* or spring equinox; and the other, where he crosses six months afterward, going south, is called the *autumnal* equinox. At these periods the days and nights are equal, and that is the reason why they are called equinoctial points, from two Latin words, *aquus,* equal, and *nox,* night. These two points are in the signs Aries (♈) and Libra (♎), and are so marked on the diagram of the zodiac.

The relative positions of the equinoctial and solstitial points and the celestial equator will be better understood from the following diagram. Imagine a hoop lying horizontally, and within this another hoop teaching the first, and with one side elevated above the other, as represented in the diagram: The horizontal hoop, marked *A B,* is the *equator;* the other, and around which the signs of the zodiac are displayed, is the *ecliptic,* or apparent path of the sun. The earth is in the center, with its equator on the same plane with the celestial equator. The equator of the earth is marked *e e.* The line *f f is* on the same plane as the ecliptic. The two other lines, one above and one below the equator of the earth, and parallel to it, are the tropics of Cancer and Capricorn, parallel with the same lines extended in the heavens, and marked as the tropics.

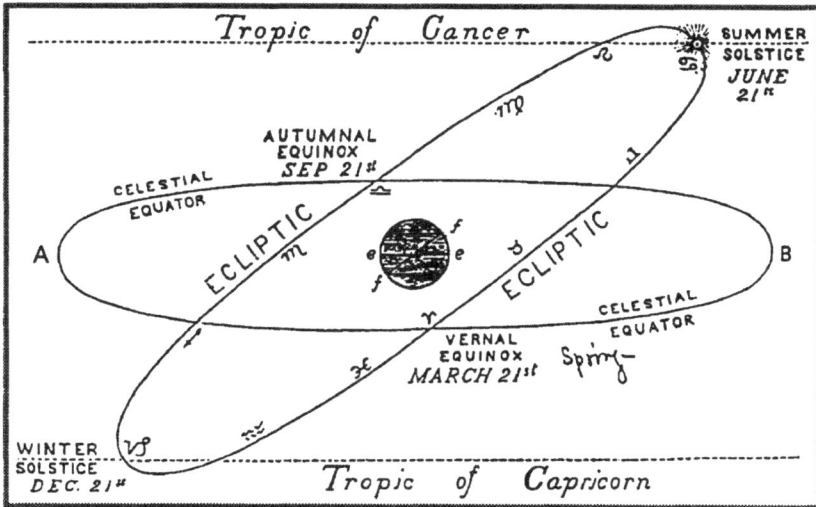

DIAGRAM OF THE ECLIPTIC AND EQUATOR

The only two points where the circle of the ecliptic and that of the equator can intersect are, of course, opposite to each other. These are the *equinoctial* points, marked Aries (♈) and Libra (♎). The *solstitial* points are those marked ♋ and ♑. Now, it is evident that when the sun leaves (♈) Aries, or the *vernal* equinox, his pathway is continually upward, and until he teaches (♋) Cancer, and there attains his highest point north of the celestial equator, as well as that of the earth *(e e)*. This is the *summer solstice.* Leaving Cancer the sun begins to decline toward the south, descending through (♌) *Leo* and (♍) *Virgo* until he reaches (♎) *Libra,* on the 23rd of September, which is the *autumnal equinox.* From this point the sun continues to descend through (♏) *Scorpio* and (♐) *Sagittarius* until the *winter solstice* is reached, at (♑) *Capricornus,* December 23rd, where the sun has reached his lowest southern declination. He is now just as far south of the celestial equator as he was north of it at the *summer solstice.*

From *Capricornus* (♑) the sun begins to ascend through (♒) *Aquarius* and (♓) *Pisces* until the *vernal equinox* is again reached. These four cardinal points, the two solstitial, together with the *vernal* and *autumnal* equinox, are therefore indicative of the four seasons of the year; for when the sun reaches the vernal equinox, *spring* begins; when he has advanced to the tropic of Cancer, the *summer* begins. His arrival at the other equinox marks the advent of *autumn;* and, when he at last reaches the tropic of Capricorn, *winter* begins.

## *The Precession of the Equinoxes*

This is the name that is given to a gradual change of place, which is constantly going on, as to the point where the sun does not cross at the same place each year; on the contrary, each time when the sun completes the circuit of the zodiac, he crosses the equator at a point a small distance *back* of the place where he did so the previous year—in consequence of which the equinoctial point is annually falling back at a uniform rate. If you will refer to the above diagram of the zodiacal points, and imagine the circle of the ecliptic being slowly turned around its center toward *Cancer*(♋), within the circle of the equator, which remains fixed—the contact between the two circles being preserved, and no change made in the angle at which they intersect each other—you will be able to gain a clear idea of the effect of this phenomenon. The point marked (♈) *Aries* would then slowly advance above the equator until the point marked ♓ was at the intersection of the two circles. The Fishes (♓) would then be on the equinox, which is now the case. In time, this motion being continued, ♒ would occupy that place, and so on.

The point where the sun crossed the equator was once in the *constellation Aries,* but in the long progress of centuries the place of the sun's crossing has fallen back 30° from the first degree of that constellation, so that the vernal equinox is now really in the constellation *Pisces,* the Fishes; or, in other words, the sun enters the stars of the *constellation* Pisces on the 21st of March, and not those of the *constellation* Aries, as it did twenty-two centuries ago, as we are informed by Hipparchus. The place, however, where the sun crosses the celestial equator has continued to be, and still is, and will continue to be, marked by the *sign* (♈) Aries, so that the *sign* of Aries now marks the place in the zodiac of the *constellation* of the Fishes. The *signs* and the constellations are therefore no longer in the same place. Hence, in order to make our chart of the zodiac (page 42) strictly correct, the *sign* Aries (♈) should be placed against the constellation Pisces, the sign ♉ against the constellation Gemini, the sign ♊ against the constellation Cancer, and so around the entire circle. It was only for the sake of simplicity and a greater ease of explanation that it was not so represented. When, therefore, it is said in astronomical language that the summer solstice is in Cancer, it is the *sign* (♋) only which is intended for the sun at that period now really enters the stars of the constellation Gemini. In like manner the winter solstice is in the *sign* ♑, but in the *constellation* Sagittarius; the autumnal equinox in the *sign* ♎, but in the *constellation* Virgo.

This precession of the equinoxes is still going on, but the other four cardinal points of the zodiac will always continue to be marked by the *signs* ♋ and ♑ and ♈ and ♎, without regard to the *constellations* which are the sun actually enters at those periods. It is by this means that astronomers are able to register upon the face of the heavens this apparent movement of the stars. This phenomenon is called the *precession* of the equinoxes, although it is really a falling back of the equinoxes, although it is really a falling back of the equinoctial point; however, as it causes the stars apparently to advance, it has been called a "precession."

The rate of this motion has been determined by long-continued observations, and is a little more than fifty and a quarter seconds $(50^1/_4$") of a degree each year. It therefore takes the equinoctial point about 2,140 years to fall back an entire sign, or 30°. In 25,791 years it will make a complete revolution of the whole circle of the zodiac. This period is called the GREAT PLATONIC YEAR, because that philosopher taught that after it had elapsed the world would begin anew.

Hipparchus, who made the first catalogue of the stars known to us, and who is called the father of astronomy, was the first who observed the motion of the stars. He informs us time, twenty-two centuries ago, the equinoctial point was in the first degree of the constellation Aries.

> The Hindoo astronomer, Varaha, says the southern solstice was certainly once in the middle of *Aslcha* (Leo); the northern in the first degree of *Dhanishta* (Aquarius). Since that time the solstitial as well as the equinoctial points have gone backward on the ecliptic 75°. This divided by $50^1/_4$", gives 5,373 years. Sir W. Jones says that Varaha lived when the solstices were in the first degrees of Cancer and Capricorn, or about four hundred years before Christ. (Burritt)

A brief description of the yearly progress of the sun will help much to the understanding of subsequent portions of this work. What follows will be better understood by again referring to the figure of the zodiac. The ancients began the year at the vernal equinox. If we start with the sun at that point and follow his progress, it will be observed that, after leaving the sign Aries (♈), in March, he next enters the signs Taurus and Gemini (♉ and ♊), and that, as he advances from the vernal equinox, the sun is daily increasing in light, heat, and magnetic power. On the 21st of June the summer solstice is reached, the summer begins. This is the longest day in the year, and the sun then attains his greatest brilliancy and dispenses the most light. All through the summer months his heat and power are at the greatest, but as he approaches the sign (♎) Libra, at the autumnal equinox, the days gradually shorten, and as he leaves Libra they grow dark and short with great rapidity. In October and November the sun enters the signs Scorpio and Sagittarius (♏ and ♐), and the cold and stormy winds begin to herald the approach of winter. The sun daily loses power, his rays grow rapidly more feeble and pallid until Capricorn (♑) is reached at the winter solstice. At this period occurs the shortest day of the year, and from that time forward the sun seems to lie dead in the cold embrace of winter, until, again approaching Aries (♈) and the vernal equinox, he begins to show symptoms of returning life. When Aries (♈) is reached, the sun begins to again manifest his power. The snow and ice melt away beneath his reviving rays, and vegetation begins to show itself.

After the vernal equinox the sun rapidly regains his vitality, and all nature with him springs from the torpidity and death of winter. The earth and the heavens, clothed once more in light and beauty, rejoice in a new life. It was this succession of phenomena, marking the yearly progress of the sun in the zodiac, that led the ancients, in their poetical and allegorical language, to represent the sun as being slain in the autumn and winter, and being restored to life again in the spring and summer.

That part of the zodiac reaching from to ♈, embracing the seasons of flowers and fruits, may well be described as the region of life, light, and beauty, while all that portion extending from the autumnal equinox through the signs ♏, ♐, ♑, to the last point of ♓, is in like manner the domain of darkness, winter, and death.

**Chapter 4**

# *WHAT THE ANCIENTS KNEW ABOUT ASTRONOMY*

IT WILL BE NECESSARY to ascertain what the ancients knew about astronomy, as what is offered for consideration in the body of this work presupposes they had made great progress in that science, not, indeed, equal to ours, but far greater than was thought to be the case before recent discoveries in Asia Minor and Egypt, or than is even now generally supposed by those who have not particularly inquired into the matter.

Rawlinson, speaking of the Chaldeans, says,

> We are formed by Simplicius that Callisthenes, who accompanied Alexander to Babylon, sent to Aristotle from that capital a series of astronomical observations, which he had found preserved there, extending back to a period of 1,903 years before Alexander's conquest of that city, or 2234 B.C.

This would be over four thousand years ago. Ideler, quoted and endorsed by Humboldt, says,

*57*

> The Chaldeans knew the mean motions of the moon with an exactness which induced the Greek astronomers to use their calculations for the foundations of a lunar theory.

Ptolemy, also, used Chaldean observations which extended back 721 B.C. Diodorus Siculus says the Chaldeans attributed comets to natural causes, and could fortell their reappearance. He states that their recorded observations of the planets were very ancient and very exact. According to Seneca, their theory of comets was quite as intelligent and correct as that of the moderns. He says they classed them with the planets, or moving stars, that had fixed orbits. (Baldwin's "Prehistoric Nations.")

The Egyptians also made great progress in astronomy, geometry, and other sciences in the time that preceded the accession of Menes, their first king, which takes us back to a period now over five thousand years ago. (Wilkinson's "Ancient Egyptians.") Herodotus informs us (Book II, Chapter IV) that the Egyptians "were the first to discover the solar year, and to portion out its course into twelve parts." They "obtained this knowledge," he says, "from the stars." The Egyptians were inventors of what we call "leap year," for they made every fourth year to consist of three hundred and sixty-six days, so as to correct and keep the calendar in order. This must have been done at least 1322 B.C., according to Wilkinson. Caesar was indebted to an Egyptian astronomer, Sosigenes, for his famous correction of the calendar. Plato ascribes the invention of geometry likewise to the Egyptians. Herodotus also says, "Geometry first came to be known in Egypt, whence it passed into Greece" (Book II, Chapter CIX). The Egyptians knew the true system of the universe. They were acquainted with the fact that the sun is the center of the solar system, and that the earth and other planets revolve about it, in fixed orbits. They knew that the earth is of a globular shape, and revolves on its own axis, thus producing day and night. They also knew of the revolution of the moon about the earth, and that the moon shines by the reflected light of the sun. They understood the calculation of eclipses; they were aware of the obliquity of the ecliptic, and that the milky-way is a collection of stars. They also seem to have understood the power of gravitation, and that the heavenly bodies are attracted to a center. (Rawlinson's "Herodotus," Appendix to Chapter VII, Book II, and authorities there quoted.) Pythagoras, who introduced the true system of the universe into Greece, received it from Oenuphis, a priest of On, in Egypt.

This great proficiency alone in astronomy would make it a matter of certainty that the ancients possessed the telescope, long supposed to be one of the grandest inventions of modern times, as the discovery of many of these astronomical facts, known to the Chaldeans and Egyptians, would simply be impossible without it. A knowledge of the heliocentric system, long lost, and only rediscovered by Copernicus, was not considered actually demonstrated or credited by the moderns until the rediscovery of the telescope, which revealed the phases of Venus, and so put the matter beyond doubt. We, however, are not left to conjecture only on this point, for there is some positive testimony that the ancients possessed the telescope. I quote again from Baldwin's "Prehistoric Nations":

> Much progress in astronomy requires the telescope, or something equivalent, and it seems necessary to believe that the ancients had such aids to eyesight. Layard and others report the discovery of a *lens* of considerable power among the ruins of Babylon. Layard says this lens was found with two glass bowls in a chamber of ruins called Nimroud. It is plano-convex, an inch and a half in diameter, and nine tenths of an inch thick. It gives a focus of four and a half inches from the plane side. Sir David Brewster says, "It was intended to be used as a lens, either for magnifying or condensing the rays of the sun."
>    (See Layard's Nineveh and Babylon," pp. 16-17, Chapter VIII)

This settles the fact that the ancients at a very remote period were familiar with all those laws of optics a knowledge of which is required to invent the telescope, and also with the manufacture of glass, so necessary for lenses designed for telescopic uses. That the art of making glass was known to the ancients—a fact once doubted—is proved also by discoveries in Egypt, where the whole process of blowing glass has been found depicted on the ancient monuments. So great was the skill of the ancient Egyptians in making vases of various colored glass, that our best European workmen of modern times cannot equal them. Glass was also one of the great exports of the Phoenicians. The Egyptians, however, surpassed all others, and some vases of brilliant colors, presented by an Egyptian priest to the Emperor Hadrian, were considered so valuable and curious that they were never used except on grand occasions. Some of the details of Egyptian glass in mosaic work (by a process common with that people more than three thousand years ago), such as the feathers of birds, so find *as to be only made out with a lens,* which means of magnifying must therefore have been known in Egypt at the remote period when this mosaic glasswork was made. This shows us that the use of the *lens* was not confined to Assyria at an early epoch, nor yet was a recent discovery there. (Wilkinson's "Ancient Egyptians")

Mr. Baldwin, in his work, continues as follows:

> Even the Greeks and Romans, with lower attainments in astronomy, had aids to eyesight. They are mentioned in "De Placitus Phil.," lib. iii, c. v, attributed to Plutarch, also in his "Vita Marcelli," and by Pliny, "Hist. Natur.," lib. xxxvii, c. v, where he says that, in his time, artificers used emeralds, to assist the eye, and that they were concave, the better to collect the visual rays.

He adds that Nero used such glasses when he watched the fights of the gladiators.

## Chapter 4. What the Ancients Knew about Astronomy

There is frequent mention of concave and convex glasses used for optical purposes, and they evidently came from Egypt and the East. Iamblichus tells us, in his life of Pythagoras, that Pythagoras sought to contrive instruments that should aid hearing as effectively as optic glasses and other contrivances aided sight. Plutarch speaks of mathematical instruments used by Archimedes "to manifest to the eye the largeness of the sun." Pythagoras and Archimedes both studied in Egypt and Phoenicia, and probably in Chaldea. Pythagoras, who lived in the sixth century before Christ, is said to have "visited Egypt and many countries of the East" in pursuit of knowledge; and Archimedes, who lived after the time of Alexander, spent much time in Egypt, "and visited many other countries."

It appears that, in the time of Pythagoras, "optic glasses," contrived to increase the power of vision, were so common as not to be regarded as objects of curiosity, and there can be no reasonable doubts that they were first invented by the *great men who created that profound science of astronomy* for which people of Cushite origin were everywhere so preeminently distinguished, and which was *so intimately connected with religion.* (Baldwin's "Prehistoric Nations," pp. 178-179)

The authorities above quoted, and the considerations advanced, render it certain that the ancients not only possessed the telescope, or its full equivalent, but also had attained a proficiency in astronomy abundantly sufficient for them to have originated the philosophical astronomical allegories ascribed to them in the course of this work. Their religion, says Mr. Baldwin, was intimately connected with astronomy.

Having thus disposed of matters which were deemed to be necessary preliminaries to our subject, the consideration of the connection between the astronomical ideas of the ancients and their religion, and the origin and true meaning of the masonic traditions, legends, symbols, and emblems, will no longer be delayed. What we have to say will be embodied in a series of questions and answers. This is a mode of instruction made familiar to all brothers of the fraternity by the masonic lectures pertaining to the various degrees. It is therefore believed that this mode will be the most acceptable to masonic readers, and not displeasing to others. It has the additional merit of permitting a degree of condensation and brevity not inconsistent with clearness of explanation, which no other method possesses.

# Part Second

*Arranged in the Form of a Masonic Lecture,
and illustrated by a Zodiacal Diagram.*

# Chapter 5

# *MASONIC ASTRONOMY*

## *Name of the Order*

Q. By what name were masons anciently known?

A. Long before the building of King Solomon's temple, masons were known as the *"Sons of Light."* Masonry was practiced by the ancients under the name of Lux (Light), or its equivalent, in the various languages of antiquity.

Q. What is said to be the origin of the word
  *"masonry"?*

A. We are informed by several distinguished writers that it is a corruption of the Greek word *mesouraneo,* which signifies "I am in the midst of heaven," alluding to the sun, which, being "in the midst of heaven," is the great source of light. Others derive it directly from the ancient Egyptian *Phre,* the sun, and *Mas,* a child: *Phre-Massen*—i.e., Children of the Sun, or, Sons of Light.

## *Astronomy and Geometry*

Q. What two sciences have always been held in especial reverence by masons?

A. Astronomy and geometry, the latter because it is the foundation of the former

## *The Lodge*

Q. How ought every lodge to be situated?

A. Due east and west.

Q. Why So?

A. Because, in the language of Dr. Hemming, a distinguished brother and masonic writer, "the sun, the glory of the Lord, rises in the east and sets in the west."

Q. What are the dimensions and covering of a lodge?

A. Its dimensions are without limit, and "its covering no less than the clouded canopy or starry-decked heavens." In the language of Oliver,

> Boundless is the extent of a mason's lodge—in height to the topmost heaven—in depth to the central abyss— in length from east to west—in breadth from north to south.

Q. How many lights has a lodge?

A. According to Dr. Oliver, in his *Dictionary of Symbolical Masonry,* a lodge has three lights—one in the east, another in the west, and another in the south.

Q. Why are they so situated?

A. Dr. Oliver, in his work just named (see page 163, "Lesser Lights"), says they are so situated "in allusion to the sun, which, rising in the east, gains the meridian in the south, and disappears in the west." These luminaries, says Dr. Oliver, in the same place, "represent, emblematically, the sun, the moon, and the master of the lodge." The same authority informs us that a lodge "has no light in the north, because the sun darts no rays from thence." (See p. 109, "Fixed Lights.")

Q. Of what is a lodge therefore emblematic?

A. The whole earth illuminated by the sun, shining from the east, south, and west; covered by day with a "clouded canopy" and at night by "the starry-decked heavens." Says Hutchinson, a standard masonic author, "The lodge, when revealed to an entering mason, discovers to him the representation of the world."

## *The Officers' Stations*

Q. Why stands the Junior Warden in the *south,* the Senior
   Warden in the *west,* and the Master in the *east?*

A. Because the sun rises in the *east* to open and govern the day, and sets in the *west* to close the labors of the same; while the sun in the *south* admonishes the weary workman of his midday meal, and calls him from labor to refreshment. Dr. Oliver informs us, in his dictionary, that

> the pedestal, with the volume of the sacred laws, is placed in the *eastern* part of the lodge, to signify that as the sun rises in the east, to open and enliven the day, so is the Worshipful Master placed in the east to open the lodge, and to employ and instruct the brethren in masonry.                    (See article "East")

Gadicke, another masonic writer, says, "The sun rises in the *east,* and in the *east* is the place for the *Worshipful Master";* and, finally, Dr. Hemming, speaking of the three principal officers of the lodge, says:

> The Sun rises in the *east* to open the day, and dispenses life and nourishment to the whole creation. This is well represented by the *Worshipful Master,* who is placed in the *east* to open the lodge, and who imparts light, knowledge, and instruction to all under his direction. When it arrives at its greatest altitude in the *south,* where its beams are most piercing, and the cool shade

most refreshing, it is then also well represented by the *Junior Warden,* who is placed in the *south* to observe its approach to meridian, and at the hour of noon to call the brethren from labor to refreshment. Still pursuing its course to the *west,* the sun at length closes the day and lulls all nature to repose; it is then fitly represented by the *Senior Warden,* who is placed in the *west* to close the lodge, by command of the Worshipful Master, after having rendered to everyone the just reward of his labor.

It is thus apparent that not only the position, form, dimensions, lights, and furniture of the lodge, but also its principal officers, their respective stations and duties there, all have reference to the sun. The several quotations made from the public and authorized writings of distinguished members of the craft render this plain to unmasonic readers. All members of the fraternity will find this fact more fully confirmed in their minds from their own knowledge of the particulars of the ritual itself.

### The Masonic Journey

Q. In what direction are masons instructed to travel?

A. Toward the east, in search of light.

Q. Why so?

A. Because the sun rises in the east, and is the great source of light.

## Masonic Words and Names

Q. What does the word of an E. A. M. signify?

A. It has more than one collateral meaning, pronounced or written either forward or backward, but if divided into the radicals of which it is composed it will be found to signify *the Fire-God* or *Quickening Fire*—i.e., the sun.

Q. What does the word F. C. M. signify?

A. This word, if divided into its radicals, means the moon.

Q. What does the word of a M. M. signify?

A. The roots of which it is composed signify the Benevolent God of Fire—i.e., the sun; and, as it was by the aid of fire that metals were first brought into a state fit for the use of man, this divinity was named Vulcan by the Romans, and worshipped by them.

Q. What does the name of O. G. M. H. A. signify?

A. It is derived from two roots, which signify the origin or manifestation of light; also he who was and is.

Q. What, then, does the whole name signify?

A. The source of eternal light—i.e., the sun—taken as an emblem of Deity.

Q. Whom, therefore, does O. G. M. H. A. represent.

A. The great source of light—the sun.

## *The Royal Arch*

Q. What is the Royal Arch?

A. It may be defined in nearly the same words as the lodge, and is no less than the starry vault of heaven, or great zodiacal arch, reaching from the vernal to autumnal equinox.

Q. How is the Royal Arch supported?

A. By three of the cardinal points of the zodiac: being the equinoctial points at the base and solstitial point at the summit.

Q. Of what are these three points emblematic?

A. Like the three pillars of the lodge, they are emblematic of WISDOM, STRENGTH, AND BEAUTY. Dr. Oliver, in his *Dictionary of Symbolical Masonry,* informs us that

> the lodge is supported by three pillars, which are Wisdom, Strength, and Beauty; because no piece of architecture can be termed perfect unless it have

wisdom to contrive, strength to support, and beauty to adorn.

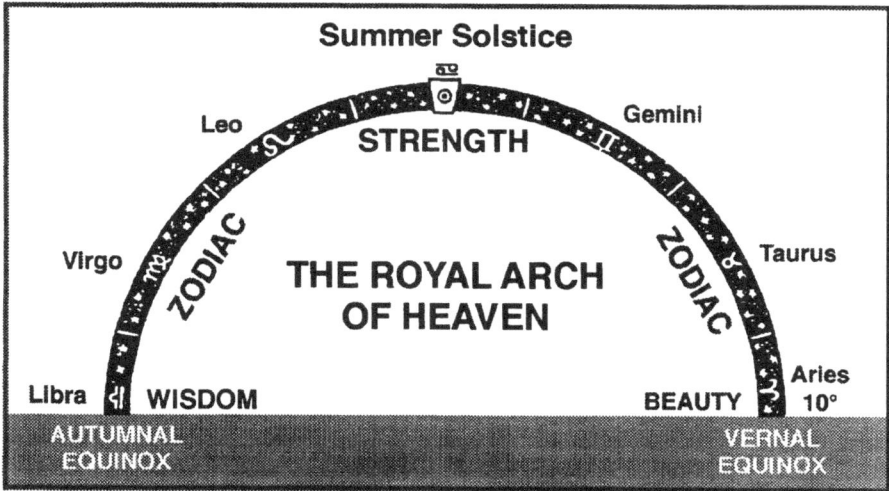

**Summer Solstice**

Leo

**STRENGTH**

Gemini

Virgo

*ZODIAC*

**THE ROYAL ARCH OF HEAVEN**

*ZODIAC*

Taurus

Libra     **WISDOM**

**BEAUTY**

Aries
10°

**AUTUMNAL EQUINOX**

**VERNAL EQUINOX**

THE ROYAL ARCH

Q. Why are the three great zodiacal points which support the Royal Arch of heaven also emblematic of wisdom, strength, and beauty?

A. At the time of the building of King Solomon's temple, or about 1004 B.C., the celestial equator cut the ecliptic at about 10° of the constellation Aries. At that period the constellation Leo was therefore near the solstitial point, and summit of the zodiacal arch. Now, as the lion is the strongest of beasts, and because the summit or key of an arch is its strongest point, and the sun, when he reaches that point, has the greatest glory and power, it being the summer solstice, when the day is the longest—that point is emblematic of *strength.* The vernal equinox signifies *beauty,* because it marks the opening of spring, which is the season of beauty, and adorns both the heavens with light and the earth with flowers. The autumnal equinox denotes *wisdom,* because it is the season of maturity. Near that point is also seen the constellation of the Serpent, in all ages typical of wisdom, and in many ancient zodiacs this point is designated by the figure of a serpent.

Q. How may the truth and beauty of this be more strongly
   impressed upon the mind?

A. By contemplating the Royal Arch itself as it actually appears in the heavens.

Q. What is required in order to be able to do so?

A. A sufficient knowledge of the constellations and a favorable time of observation.

Q. What is the most favorable time to observe the Royal Arch?

A. If we wish to observe the constellations as they were at the summer solstice at the time of the building of King Solomon's temple, we should view the heavens about the 1st of August, but as the sun in the south at high twelve, by its overpowering light, prevents the proper stars being seen, it will be necessary to defer our observations for six months, or until about the 5th of February, at which time the same stars are visible at *midnight.* "Low twelve," about the 5th of February, is, therefore, the best time to view the Royal Arch.

If we then take our station, looking south, and lift our eyes to the vast arch of heaven, the spectacle will be one of unsurpassed magnificence, and to an intelligent mason eloquent with the truths of his profession. Far up the blue concave, and within less than 30° of the summit of the arch, will be seen the constellation *Leo,* typical of STRENGTH; on either side will be seen the constellations *Aries* and *Libra,* which anciently marked the equinoctial points, and upon which the whole majestic arch seems to rest.

*Libra,* the Balance, is typical of what WISDOM which, in the scales of Reason, duly weighs and considers all things; while *Aries,* marking the ancient place of the vernal equinox, is typical of BEAUTY, and also gives a sure token that the sun, which lies dead in the cold arms of Night and Winter, will arise again in the springtime, clothed with new life and power. The vernal equinox, or sign *Aries,* is therefore also the symbol of immortality, and teaches that the soul of man will rise in glory from the darkness of the grave. It also reminds masons of the lamb, "which has in all ages been considered an emblem of innocence," and admonishes him

> of that purity of life and conduct which is so essentially necessary to his gaining admission into the celestial lodge above, where the Supreme Architect of the universe presides.

In the east, in close proximity to Libra, stands the beautiful virgin of the zodiac, the constellation *Virgo.* In her left hand gleams the bright star *Spica,* while not far away toward the north *Arcturus* shines in splendor. In the west *Taurus* is seen with the *Pleiades. Orion* also lifts his giant form along the sky, sublime in his majesty and beauty. Still lower down, and near the horizon, blazes the great sun-star *Sirius. Procyon* also shines with almost equal glory higher up the sky.

*Gemini,* too, the twin brothers *Castor* and *Pollux,* offspring of the mighty Jove, adorn the heavens. In the north *"Cassiopea* sits in her golden chair," while the *Great Bear* guards the pole. There, too, are seen *Cepheus,* and *Andromeda* bound to the rock with chains. The polar star, emblem of eternal constancy, shines with a steady light; while around the pole the scaly *Dragon* coils his glittering folds. Meanwhile, as we continue to observe the midnight meridian, other constellations, as they rise, light up the gleaming arch, each teaching a different lesson, and *all*—

> "Forever singing, as they shine,
> The hand that made us is divine."

The accompanying diagram of the Royal Arch is but a geometrical projection, and, therefore, gives nothing more than the relative positions of the various constellations and signs of the Royal Arch. The summer solstice is represented as the *key-stone* of the arch, and has the astronomical sign of the sun inscribed upon it, showing that on the 21st of June the sun is exalted to the summit of the arch. It was formerly thought that the ancient Egyptians were not acquainted with "arch" in architecture, but late discoveries show that it was known to them at lest 2100 B.C. (See Wilkinson's "Egyptians of the Time of the Pharaohs," p. 137.)

## King Solomon's Temple

Q. Of what was King Solomon's temple emblematic?

A. That temple not made with hands, eternal in the heavens.

Q. Has the word "temple" any meaning significant of this?

A. All ancient temples were originally dedicated to the worship of the sun and the summer celestial orbs, whose circuit in the heavens each year was emblematic represented in the details of their construction and ornaments. The word "temple" is from *tempus,* time; *templum* comes from *tempus,* and the word "temple" is therefore synonymous with *tempus,* time, or the year.

Q. By whom is time—i.e., the temple—each year beautified and adorned?

A. By the sun, who, from March to October, is continually engaged in beautifying the heavens and the earth.

Q. When the building of the temple commenced?

A. On the 2nd day of Zif, or about the 21st of April.

Q. When was the temple finished?

A. On the 4th day of Bul, or about the 21st of October.

Q. Have those dates any astronomical significance?

A. They have. On the 21st of April the sun enters *Taurus,* and the plowing and planting begin. On the 21st of October the sun enters *Scorpio;* "the summer is over and the harvest is finished." It was, therefore, that the sun, in his passage through the *seven* signs (typical of years), from Aries to Scorpio, was said, emblematically, to raise the Royal Arch, beautify and adorn the heavens, and bring forth the bountiful fruits of the earth.

Q. Is it, therefore, to be understood that the whole account of the building of King Solomon's temple, as given in the masonic tradition, is an astronomical myth?

A. By no means, for there is no fact more certain than the building of King Solomon's temple, as both sacred and profane history testify. It is nevertheless true that the masonic tradition respecting it is one of mystical import. It contains within itself not only the history in part of the building of an actual earthly and material temple, but also an emblematic description of the heavens and the earth, as well as of the particulars of the annual passage of the sun among the twelve signs of the zodiac. There is also good reason for believing the temple itself was expressly built, so as to be in its various parts emblematic of the whole order of nature.

Josephus (most learned of Jews) directly informs us that the *tabernacle,* which was a prototype of the temple, was thus emblematic in its construction. He says, speaking of the tabernacle and vestments of the high-priest, that,

> if anyone, without prejudice and with judgment, look upon these things, he will find *they were every one made* in way of *imitation and representation of the universe.* When Moses distinguished the tabernacle into three parts, and allowed two of them to the priests, as a place accessible and common, he denoted the land and the sea, these being of general access to all; but he set apart the third division for God, because heaven is inaccessible to men. And when he ordered twelve loaves to be set upon the table, he denoted the year as distinguished into so many months. By branching out the candlestick into seventy parts, he secretly intimated the *Decani,* or seventy divisions of the planets; and as to

the seven lamps upon the candlesticks, they referred to the course of the planets, of which that is the number. The veils, too, which were composed of four things, they declare the four elements; for the fine linen was proper to signify the earth, because the flax grows out of the earth. The purple signifies the sea, because that color is dyed by the blood of a sea shellfish; the blue is fit to signify the air, and the scarlet will naturally be an indication of fire. Now, the vestment of the high priest, being made of linen, signified the earth; the blue denotes the sky, being like lightening in its pomegranates, and in the noise of the bells resembling thunder. And for the ephod, it showed that God had made the universe of four (elements); and as for the gold interwoven, I suppose it related to the splendor, by which all things are enlightened. He also appointed the breastplate to be placed in the middle of the ephod to resemble the earth, for that was the very middle place of the world. And the girdle which encompassed the high priest round signified the ocean, for that goes round about, and includes the universe. Each of the sardonyxes declares to us the sun and moon—those, I mean, which were in the nature of buttons on the high priest's shoulders. And as for the twelve stones, whether we understand by them the months, or whether we understand the like number of the signs of that circle which the Greeks call the zodiac, we shall not be mistaken in their meaning. As for the mitre, which was of blue color, it seems to me it means heaven, for how otherwise could the name of God be inscribed upon it? That it was also illustrated with a crown, and that of gold also, is because of that splendor with which God is pleased. Let this explanation suffice at present.      ("Antiquities," Book III, Chapter VII, 7)

The concluding sentence of this quotation conveys a clear intimation that many other emblematic particulars in the construction of the tabernacle might be pointed out. Now, as the "holy place," and veils, candlesticks, lamps, vestments, and other particulars of the tabernacle were specifically reproduced in the temple, we may safely conclude that the temple itself was so built as to be also emblematic, in its several parts, of the universe. Nor when we reflect that the designs for the temple, as well as the tabernacle, are said to have been given by God himself, need we be surprised at this, for what more reasonable than to suppose that, when the great Creator of all things revealed the designs for a temple to be dedicated to himself, it should thus be made in all its parts emblematic of the sum of all his other works—the entire universe? The lodge, according to all masonic writers, is emblematic of King Solomon's temple; it is therefore easy to see why it is also emblematic of the heavens and the earth. It could not be the one without being also the other. It also naturally follows that the masonic tradition is thus possessed for a threefold character:

1. It is in part an actual history of the building of King Solomon's temple.
2. It is an emblematic description of the heavens and the earth.
3. By a system of allegorical and astronomical symbols it is the depository of a high code of morals.

In its triune aspect it is, therefore, HISTORICAL, SCIENTIFIC and MORAL. In it the two accounts of the building of the actual and the mystical temple, the earthly and the heavenly one, are curiously interwoven and permeate each other. Yet, the astronomical key being given, they may be separated, and each contemplated by itself.

## *Hiram Abif*

In them Hiram Abif appears both as an authentic and a mystical personage, he is not only the cunning craftsman employed by King Solomon to beautify and adorn the actual temple, but also an emblematic being, representing the sun, who, by his magnetic power, raises the Royal Arch of heaven, and beautifies and adorns the terrestrial and celestial spheres, for which reason his name has a twofold meaning, significant of both characters.

It is also true that to some extent the life and conduct of the real personage is emblematic of the mystical one, yet they differ in several important particulars:

1. The *mystical* Hiram is represented in the masonic tradition as being an architect, superintending the building and drawing out the plans for the temple.

The *real* Hiram, as mention in history, was, according to the Bible, and also Josephus, no architect at all, and drew out none of the designs for the temple.

2. The *mystical* Hiram, according to masonic tradition, is represented as having lost his life suddenly before the completion of the temple, in the midst of his labors, and with many of his designs unfinished.

On the contrary, the *historical* Hiram, as we are expressly informed in the sacred Scriptures, lived to finish all his labors in and about the temple, and for King Solomon.

For the benefit of unmasonic readers, we will give the substance of the masonic tradition relating to Hiram Abiff, which is taken word for word from Dr. Oliver's *Dictionary of Symbolical Masonry,* a work authorized by the highest masonic bodies in England and America. Says Dr. Oliver:

We have an old tradition delivered down, orally, that it was the duty of Hiram Abiff to superintend the workmen, and that the reports of the officers were always examined with the most scrupulous exactness. At the opening of the day, when the sun was rising in the east, it was his constant custom, before the commencement of labor, to enter the temple and offer up his prayers to Jehovah for a blessing on the work. And in like manner, when the sun set in the west and the labors of the day were closed, and the workmen had departed, he returned his thanks to the Great Architect of the universe for the harmonious protection of the day. Not content with this devout expression of his feelings, morning and evening, he always went into the temple at the hour of high twelve, when the men were called from labor to refreshment, to inspect the progress of the work, to draw fresh designs upon the tracing-board, if such were necessary, and to perform other scientific labors, not forgetting to consecrate his duties by solemn prayer. These religious customs were faithfully performed for the first six years in the secret recesses of his lodge, and for the last year in the precincts of the most holy place. At length, on the very day appointed for celebrating the cap-stone of the building, he retired as usual, according to our tradition, at the hour of high twelve, *and did not return alive.*

(See article "High Twelve")

Some further particulars of the masonic legend are given in the same book, under the article "Burial-Place."

"The burial-place," says Dr. Oliver,

of a master mason, is under the Holy of Holies, with the following legend delineated on the monument: A virgin weeping over a broken column, with a book open before her. In her right hand a sprig of cassia, in her left an urn. Time standing behind her with his hands enfolded in the ringlets of her hair. The weeping virgin denotes the unfinished state of the temple; the broken column, that one of the principal supporters of masonry (H. A. B.) had fallen; the open book implies that his memory is recorded in every mason's heart; the sprig of cassia refers to the discovery of his remains; and the urn shows that his ashes have been carefully collected; and Time, standing behind her, implies that time, patients, and perseverance will accomplish all things.

Dr. Oliver also, in his ninth lecture, on the "Theocratic Philosophy of Freemasonry," speaking of Hiram Abif, says:

The legend of his death it will be unnecessary to repeat, but there are some circumstances connected with it which may be interesting. His illustrious consort, whose memory is dear to every true mason, was so sincerely attached to him that, at his death, she became inconsolable, and, refusing to be comforted, she spent the greater part of her time in lamentation and mourning over the tomb which contained his venerated ashes. The monument erected to his memory was particularly splendid, having been curiously constructed of black and white marble, from plans furnished by the Grand Warden, on the purest masonic principles, and occupied an honorable situation in the private garden belonging to the royal palace.

The foregoing authorized publication of the main facts of the masonic legend respecting the death of Hiram Abiff, contains all the particulars necessary for the illustration of our subject to unmasonic readers. To members of the fraternity, all the details of the tragic tradition are of course familiar, and many things designedly made obscure to all others will be clear to them.

The masonic tradition respecting Hiram, it will thus be seen, speaks of him as being the chief architect of the temple, superintending the workmen and drawing out designs for the construction of the temple.

The *historical* Hiram, mentioned in the Bible and by Josephus, is a different personage from the traditional one. That Hiram, who was actually sent to King Solomon, had nothing to do with furnishing the designs of the temple. We are expressly informed that the designs, form, and dimensions of the temple were all given by divine command (2 Chronicles 3). To have altered or modified them in the smallest particular would therefore have been a sin, which would have called down the instant and terrible displeasure and punishment of Jehovah. Hiram is nowhere mentioned or described in the Bible as being an architect, or even a builder. In 1 Kings 7:14, he is described as being "filled with wisdom, and understanding and cunning *to work all works in brass!*" In 2 Chronicles 2:14, the *father* of Hiram is described as

skillful to work in gold, and in silver, in brass, in iron, in stone, and in timber, in purple, in blue, and in fine linen, and in crimson; also to grave any manner of graving.

From this it is evident that the father of Hiram, who was a man of Tyre, was by profession a decorative artist and sculpture. It is probable that Hiram followed the profession of his father, according to the custom of the times, otherwise Hiram, King of Tyre, would not have thus particularly spoken of the profession of his father in describing the accomplishments of Hiram Abiff himself. King Hiram speaks of Hiram Abiff simply as "a cunning man, endued with understanding" (verse 13). Josephus also mentions Hiram, and uses the following language respecting him:

> This man was skillful in all sorts of work, but his *chief skill* lay in *working in gold, silver and brass,* by whom were made all the *mechanical works* about the temple, according to the will of Solomon.
>                         ("Antiquities," Book VIII, Chapter III, 4)

Not a word about his having anything to do with the building of the temple itself. But, as if to put this question of the temple itself. But, as if to put this question to rest, not only Josephus, but the Bible also, mentions just what these "mechanical works" were. In 1 Kings 7, is a complete list and description of them, and of all the works done about the temple and for its use by Hiram. This list of the works of Hiram is also given in 2 Chronicles 4:11-19. The same list is also given by Josephus. From these authorities we learn that Hiram made for King Solomon—

*The two pillars of brass, called Jachin and Boaz, together with their ornaments.*

*The molten sea of brass, with twelve oxen under it;* a work of great artistic beauty, but calling for the genius of a Benvenuto Cellini, rather than of a Sir Christopher Wren.

Also, *ten brazen lavers and their bases,* and many *pots, shovels, flesh-hooks,* and other altar-furniture, to be used in and about the sacrifices.

All of the foregoing articles were made of bright brass, and they were cast in clay molds, in the plains of Jordan, *between*

*Succoth and Zaredathah* (2 Chronicles 4; 1 Kings 7:45-46). *Succoth* means "booths," and was so named because Jacob built him a house there, and "made booths for his cattle" (Gen. 33:17). It *is fifty miles,* at least, in an *airline,* north by east of Jerusalem, beyond Jordan, between Peniel, near the ford of the torrent Jabbok and Shechem; while *Zaredathah,* or *Zarthan,* as it is called in Kings, is still farther north than Succoth. The words "*between* Succoth and Zaredathah," therefore, denotes that the place where the brass foundries were situated and these castings were made, was yet farther from Jerusalem than Succoth. The modern name of the torrent Jabbok is *Wady Zurka.* (See Smith's *Dictionary of the Bible,* and maps of the Holy Land at the time of David and Christ.)

As the distance in an *airline* from Jerusalem to Succoth was at least fifty miles, it is to be presumed that the distance by the traveled route was considerably more. It may be said that the clay only was procured at this distant place (distant when we consider the slow means of travel in those days), and that it was brought to Jerusalem, to be there used by the artist in making the molds for his castings. But the sacred text expressly says that the casting was done on the spot.

The scene of the labors of Hiram must, therefore, have been considerably over fifty miles from Jerusalem, or more than two days' journey, at the smallest calculation; twenty miles being an ordinary day's journey in those times and that country. Smith, in his *Bible Dictionary,* says fifteen.

Besides this, the making of the molds and patterns for them would require the personal attention if not labor of Hiram himself. The casting of large pieces, such as were required for the brazen sea, the lavers and their bases, and the pillars Jachin and Boaz, which were eighteen cubits, or about thirty-two feet in height, must have demanded his constant care and watchful attention, (see Cellini's account of the casting of his bronze Perseus, "Memoirs," vol. ii, c. xli.) These facts, taken in connection with the great number of different pieces of work, render it evident that Hiram must have been kept the greater part of his time at the distant scene of his labors, where the clay required could alone be found. It is impossible, under the circumstances, that he could have visited the temple in Jerusalem, from fifty to sixty miles distant, three times a day, or even once a day, during the seven years that the temple was being built.

Besides these works in brass, we are told that Hiram made for Solomon of pure gold *ten candlesticks for the oracle, with flowers, lamps, and tongs;* also *bowls, snuffers, basins, and censors, and hinges of gold,* for the holy place and for the doors of the temple. All this work, it will be seen, is that of a "cunning worker in metals" and a decorative artist, none of it that of an architect or builder.

The other decorative works done in and on the temple proper, consisting of carvings on the walls of figures of cherubim and palm-trees, also the golden cherubim which were set up in the holy place, are not any of them including in the list of the works of Hiram, nor, indeed, named in the same chapter.

The *mystical* Hiram of the masonic tradition, we are also told, met with a sudden death, the particulars of which are known to all members of the fraternity, before the completion of the temple. Had any such accident befallen the actual Hiram (leading, as we are told, to the suicide, from grief, of his wife), certainly the impotence of the tragic event, and the consequent delay and confusion it would naturally cause, would have led to its being recorded either in Kings or Chronicles, or both of them, but no such occurrence is anywhere mentioned in the sacred narrative, which, respecting the building and dedication of the temple, is particular and minute; nor does Josephus mention any such event. This negative testimony is almost conclusive, but we are not left to rely on that alone, for both in Kings and Chronicles we are directly informed that the *historical* Hiram, unlike the *mystical* one the masonic tradition, lived to finish all his labors. We read in 2 Chronicles 4,

"So *Hiram made an end of doing all the work* that he had made King Solomon *for the house of the Lord.*"

After the temple was finished we are told that Solomon built him a house for himself, which was, like the temple, splendidly ornamented by decorations and carvings in gold, silver, and wood. Mention is also made in Chronicles of a magnificent ivory throne, surrounded by carved figures or statutes of lions. The building and ornamentation of this house occupied *thirteen years* after the temple was finished (1 Kings 7:1). Now if Hiram was also employed by the king to decorate his own house, he must have lived at least thirteen years after the completion of the temple. That Hiram was also employed about the "kings house" is almost a certainty; for, although the list of his works, as given, makes no mention of the ivory throne, the lions, or any work done for the "king's house," yet as that list professes to be a list only of the work done by Hiram for the temple (see verse 40, also 2 Chronicles 4:11), we have no right to expect to find it including any of the other work of the artist done for the place of Solomon. The fourteenth verse of the seventh chapter of 1 Kings directly says that Hiram "wrought all of King Solomon's work." Besides this, the seven years occupied in building the temple and the thirteen in building the king's house make up the whole *twenty* years of the *contract* which Solomon had with the King of Tyre for materials and *skilled* workmen, the principal among whom was Hiram, the great artist and sculptor; and it becomes an almost conclusive presumption that Solomon kept him and the other skilled workmen the whole twenty years during which he required their aid.

As to the nature of this contract of King Solomon's with Hiram, King of Tyre, see 1 Kings 5; 2 Chronicles 2; as to its duration being twenty years, see 1 Kings 9:10; and Josephus on both points. The proof is therefore *positive* that Hiram lived to finish all his labors in and about the temple, and also highly presumptive that he continued his labors for King Solomon thirteen years afterward.

It is also just as clearly proved by history, both sacred and profane, that he the chief architect of, and furnished no designs for, the temple. According to holy writ, the designs for the temple were not only furnished by God himself, but the whole work was directed by the inspiration of the great Architect of the universe. If, then, the *historical Hiram* was no architect, but a decorative artist and sculptor only, and was not called upon to suffer a sudden death before the completion of the temple, it follows, therefore, that it is the *mystical* Hiram—representing the sun—who meets with that sad fate under the completion of the emblematic temple, and not the real one. The claim that the masonic tradition is historically true in all respects cannot be maintained, as it is in most of its main features in direct conflict with holy writ. If, however, we consider it in its allegorical character, as our ancient brethren no doubt did, if we regard it in its twofold nature, as being in part emblematic as well as historical, as before explained, all difficulties at once vanish. The entire integrity of the masonic tradition is thus fully maintained. The whole legend not only becomes the venerated depository of the most sublime astronomical facts, but is illuminated by a twofold beauty and truth.

The answer to the last question has of necessity been a somewhat lengthy one. Having disposed of it, let us renew our explanation of the astronomical allegories of the masonic tradition where we left off.

**Chapter 6**

# ASTRONOMICAL ALLEGORY OF THE DEATH AND RESURRECTION OF THE SUN

Q. Explain more fully in what manner the sun is said
   by an astronomical allegory to be slain.

A. According to all the ancient astronomical legends, the sun is said to be slain by the three autumnal months—September, October, and November, represented as assaulting him in succession.

Q. When is the sun said to be slain?

A. Near the completion of the temple, as before explained.

Q. Explain more fully by whom, and how the sun is
   said to be slain.

A. The sun is slain by September, October, and November, or the three autumnal signs, ♎, ♏, and ♐, anciently ♏, ♐, and ♑, whom he encounters in succession in his passage around the zodiac toward the winter solstice, or "southern gate of the zodiac"; so-called in the poetical language of the old Greeks, because at that point the sun has reached his lowest southern declination. The summer sun, glowing with light and heat as he reaches the autumnal equinox, enters *Libra* on the 21st of September. All through that month, and until the 21st of October, he declines in light and heat, but emerges from *Libra* (♎) without any serious harm from the attack of September. The assault of October is far more serious; and the sun when he *leaves* the venomous sign of the *Scorpion* (♏), on the 21st of November, is deprived of the greater part of his power and shorn of more than half his glory. He continues his way toward the southern tropic, and in November encounters the deadly dart of *Sagittarius* (♐), which proves fatal; for when the sun *leaves* the *third* autumnal sign, on the 23rd of December, he lies dead at the winter solstice.

Q. Why is the third attack, or that the November, said to be more fatal than that of September or October?

A. Because when the sun emerges from under the dominion of Sagittarius, the ruling sign of November, on the 23rd of December, he enters *Capricorn,* and reaches his lowest declination. That is the shortest day of the year.

In June, at the summer solstice, the bright and glorious days were over fifteen hours long. Now the pale sun rises above the gloomy horizon of December but a little more than half as long, and his feeble rays can hardly penetrate the dark and stormy clouds that obscure the sky. The sun now seems to be quite overcome by "the sharpness of the winter of death." Amid the universal mortality that reigns in the vegetable kingdom, the sun, deprived of light, heat, and power, appears dead also.

Q. Does the ancient art of *astrology* throw any further light upon this subject?

A. This science was much cultivated by the ancients under the name of the "divine art." According to the teachings of astrology, Capricorn was the "house of Saturn," the most evil and wicked in his influence of all the planets. He is called the "great infortune," and all that part of the zodiac within the signs of Capricornus and Aquarius was under his dominion. Saturn was also known as *Kronos,* or Time, which destroys all things; and, in the poetical and allegorical language of mythology, devours even his own children. The figure of Saturn with his scythe is to this day an emblem of decay and death. The sun, therefore, when he entered Capricorn, passed into the house and under the dominion of Saturn, or Death.

Q. After the sun is slain, what in allegorical language, is said to become of the body?

A. It is carried a westerly course, at night, by the three wintry signs.

Q. Why so?

A. Because, as the sun continues his course in the zodiac, he appears to be carried west by the wintry signs. This seems to be done at night, because the sun then being invisible, his change of position is only discovered by the stars which precede his rise at daybreak.

Q. What disposition is finally made of the body?

A. As it seemingly buried beneath the withered fruits and flowers—the "rubbish" of the dead vegetation of summer—in the midst of which, however, yet blooms the hardy *evergreen,* emblematic of the vernal equinox, giving a sure token that the sun will yet arise from the cold embrace of winter and regain all his former power and glory.

Q. What follows?

A. According to the Egyptian sacred legend of the death of OSIRIS, the goddess Isis ransacks the whole four quarters of the earth in search of his body, which she finally discovers indirectly *by the aid of a certain plant or shrub,* and causes it to be regularly buried, with sacred rites and great honor. According to the legend of Hiram, it was twelve fellow-crafts—emblematically representing the three eastern, three western, three norther, and three southern signs of the zodiac—who made the search of the body. It was somewhere among the twelve constellations that the lost sun was certainly to be found.

Q. By whom was the body found?

A. By *Aries* (♈), one of the three western signs, typical of those who pursued a westerly course. In going from the winter solstice to the vernal equinox, we of necessity pass *Aquarius* (♒), the Waterman, who was also known as a fisherman and a seafaring man.

Q. Where was the body found?

A. At the vernal equinox, typical of the "brown of a hill." As we pass from the winter solstice in *Capricorn* to the vernal equinox, we are constantly climbing upward; this point is therefore emblematic of the brow of a hill, and there also blooms the evergreen, typical of the approaching spring and return of nature to life.

The following is a poetical version of the foregoing portion of the solar allegory:

*A Masonic Allegory*

### Part I — The Death of the Sun.

WHEN down the zodiacal arch
♋ The summer sun resumes his march,
Descending from the summit high
With eager step he hastens by
♌ The "lordly lion" of July
And clasps the virgin in his arms.

Through all the golden August days
The sun the ardent lover plays,

♍ A captive to her dazzling charms.
But when the harvest time is o'er,

When they gathered grapes perfume the air
And ruddy wine begins to pour,
The god resumes his way once more;
And, weeping in her wild despair,
He leaves the royal virgin there.
What cares he now for Virgo's woes,
As down the starry path he goes
With scornful step, until, at last,
The equinoctial gate is passed?

Two misty columns black with storms,
While overhead there hangs between
A lurid thunder cloud, which forms
The frowning archway of the gate—
♎ The gloomy equinoctial gate,
An evil place for travelers late,
Where envious *Libra* lurks unseen;
And near the portal lies in wait
*September,* filled with deadly hate.

With stately step the god draws nigh,
Yet, such is his majestic mien,
That whether he shall strike or fly,
The trembling ruffian hardly knows,
As Phoebus through the gateway goes.

But, as the shining form came near,
The wretch's hate subdued his fear,
And, nerving up his arm at length,
He aimed a blow with all his strength
Full at the god as he went by.
In anger Phoebus turned his head—
Away the trembling coward fled.

The god, though smarting with the blow,
Disdains to follow up his foe;
And down the zodiacal path
Pursues his gloomy way in wrath.

Still blacker turn the autumn skies,
And red *Antares,* evil star,
Points out the place, more fatal far,
Where fell *October* ambushed lies.
The SUN, as if he scorned his foes,
♏ In pride and glory onward goes.

## Chapter 6. Allegory of the Death of the Sun

Not he from deadly *Scorpio* flies,
Nor pauses he, nor backward turns,
Though redder yet *Antares* burns,
And darker yet his pathway grows.

Meanwhile *October,* from his lair,
On Phoebus rushes unaware,
His murderous purpose now confessed,
And smites the sun-god in the breast.
A ghastly wound the villain makes—
With horrid joy his weapon shakes;
And, as he sees the god depart,
His hand upon his bosom pressed,
Believes the blow has reached the heart.

Along his way the sun-god goes,
Unmindful where the path may lead,
While from his breast the life-blood flows.
   The clouds around him gather now,
The crown of light fades from his brow.
♐ And soon, advancing 'mid the night,
The *Archer* on his pallid steed,
With bended bow, appears in sight.
   *November,* bolder than the rest,
Hides not behind the gloomy west;
But, striding right across the path,
Defies the god and scorns his wrath;
And, raising high his frowning crest,
These haughty words to him addressed:
"*September* and *October,* both,
You have escaped and still survive;
But I have sworn a deadly oath,
By me you cannot pass alive.
That which I promise I perform.
For I am he who, 'mid the storm,
Rides on the pallid horse of death."

While even thus the spectre spoke,
He drew his arrow to the head—
The god received the fatal stroke,
And at the *Archer's* feet fell dead.

Soon as the sun's expiring breath
Had vanished in the ether dim,
♑ *December* came and looked on him;

And looking, not a word he saith,
But o'er the dead doth gently throw
A spangled winding sheet of snow.

And when the winding sheet was placed,
♒ Comes evil *Janus,* double-faced,
A monster like those seen in sleep.
　　　An old *"seafaring man"* is he,
As many others understand,
Who carries water from the deep
And pours it out upon the land.

Now *February* next appears,
With frozen locks and icy tears,
A specter cruel, cold, and dumb,
From polar regions newly come.
These *three* by turns the body bear
At night along the west, to where
A flickering gleam above the snows
A dim electric radiance throws,
A nebular magnetic light,
Which, flashing upward through the night,
Reveals the *vernal equinox,*
And him whose potent spell unlocks
The gates of spring.
　　　　　　　　　An evergreen
Close by this spot is blooming seen.
'Tis there they halt amid the snow—
Unlawful 'tis to go farther go—
And, having left their burden there,
They vanish in the midnight air.

Yet on this very night next year
Will this same *evil three* appear,
And bring along amid the gloom
Another body for the tomb.
But still the *evergreen* shall wave
Above the dark and dismal grave,
For ever there a token sure
That, long as Nature shall endure,
Despite of all the wicked powers
That rule the wintry midnight hours,
The sun shall from the grave arise,
And tread again the summer skies.

The foregoing allegory may be fully illustrated by the figure of the zodiac on page 92. Place the image of the sun—which is on the white circle—at the summer solstice, then turn the circle slowly around toward the autumnal equinox, so that the image of the sun will pass successively by ♋, ♌, ♍, ♎, ♏, ♐, and so on until the vernal equinox is reached.

## *The Raising of Osiris, an Allegory of the Resurrection of the Sun*

Q. By what means and by whom was the sun released from the grave of winter, and finally restored to life and power?

A. By the vernal signs *Taurus* (♉) and *Gemini* (♊), and the first summer one, *Cancer* (♋), aided by the second one, *Leo* (♌); or, in other words, by April, May, and June, aided by July.

Q. Explain this more fully.

A. When the sun arrives at the *vernal equinox,* he first gives unequivocal tokens of a return to life and power. In April he enters *Taurus* (♉), and in May *Gemini* (♊). During these two months he greatly revives in light and heat, and the days rapidly lengthen. The sun, however, does not attain the summit of the zodiacal arch until the *summer solstice,* in June, when he enters Cancer (♋), the first summer sign and the *third* from the vernal equinox. Nor does he regain all of his energy and power until he enters Leo (♌) in July.

On the 21st of June, when the sun arrives at the summer solstice, the constellation *Leo*—being but 30° in advance of the sun—appears to be leading the way and to aid by his powerful paw in lifting the sun up to the summit of the zodiacal arch. April and May are therefore said to fail in their attempt to raise the sun; June alone succeeds by the aid of *Leo.* When, at a more remote period, the summer solstice was *in Leo,* and the sun actually entered the stars of that constellation was more intimate, and the allegory still more perfect.

ASTRO-
MASONIC
EMBLEM ----
SUN IN LEO

This *visible* connection between the constellation *Leo* and the return of the sun to his place of power and glory, at the summit of the Royal Arch of heaven, was the principal reason why that constellation was held in such high esteem and reverence by the ancients. The astrologers distinguished *Leo* as the "sole house of the sun," and taught that the world was created when the sun was in that sign.

> The lion was adored in the East and the West by the Egyptians and the Mexicans. The chief Druid of Britain was styled a lion. The national banner of the ancient Persians bore the device of the sun in Leo. A lion couchant with the sun rising at his back was sculptured on their palaces.
>
> ("Signs and Symbols" of Dr. Oliver, who seems, however, to have entirely overlooked the true reason for this widespread adoration of the lion.)

The ancient device of the Persians is an astronomical allegory. It might well be adopted as an astro-masonic emblem by us.

After the sun leaves Leo, the days begin to grow unequivocally shorter as the sun declines toward the autumnal equinox, to be again slain by the *three* autumnal months, lie dead through the *three* winter ones, and be raised again by the *three* vernal ones. Each year the great tragedy is repeated, and the glorious resurrection takes place.

Thus, as long as this allegory is remembered, the leading truths of astronomy will be perpetuated, and the sublime doctrine of the immortal nature of man, and other great moral lessons they are thus made to teach, will be illustrated and preserved.

The diagram on page 92 is intended, by a figure of the zodiac, to illustrate the yearly progress of the sun among the twelve signs, with especial reference to the allegory of his death and return to life, as explained in the preceding pages. In this figure of the zodiac the vernal equinox is represented as being somewhere between the constellations *Aries* and *Taurus,* and the summer solstice between Cancer and Leo. Such was the case at the period of the building of King Solomon's temple, and for a long period before that; only, the farther back we go in time, the nearer Leo will be to the summer solstice, in consequence of the precession of the equinoxes, as has been explained in a preceding chapter.

In order to fully illustrate the allegory by means of the diagram, bring the image of the sun, on the *white* circle, to the summer solstice, immediately under the key-stone, and figure of the personified sun-god, at the top of the *grey* circle; then slowly turn the *white* circle toward the autumnal equinox, so that the image of the sun in the *white* circle will pass successively by the constellations from *Leo* to the winter solstice at the bottom of the grey circle. This closes the first part of the allegory. Continue to turn the white circle until the vernal equinox is reached, and then on through Taurus, Gemini, and Cancer ( ♉, ♊, and ♋), until the point of the sun's exaltation is once more attained. This will give a correct representation of the annual passage of the sun among the twelve signs of the zodiac as it actually appears in nature, and also illustrate the whole course of the solar allegory.

The following is a poetic version of the second part of the solar allegory:

## *A Masonic Allegory*
### Part II — The Resurrection of the Sun

IN silence with averted head
by night the *"evil three" have* fled.
And cold and stiff the body lies
Beneath the gloomy winter skies.
    Yet, had you been a watcher there.
That dismal night beside the dead.
Had you that night been kneeling there,
Beside the dead in tears and prayer.
You might have seen, amid the air,
A flickering, dim, auroral light,
Which hovered on the midnight air,
And, seeing in the gloomy sky
This mystic strange, celestial light
Contending with the powers of night.
You might have taken hope thereby.

There was, alas! no watcher there
To mark this radiance in the air.
To gaze with ernest, tearful eye
Upon this radiance in the sky.
There was no watcher there, alas!
To ask in anxious whispers low,
"Will not this light still brighter grow,
Or will it from the heavens pass
And leave me plunged in deeper gloom
Beside this cold and lonely tomb?"

Meanwhile the light increased—although
Beside the grave no mourner stood
Amid the lonesome solitude—
And as with tints of blue and gold,
And flashes of prismatic flame,
It lighted up the midnight cold,
Along the plain in beauty came
A shining and majestic form,

And as it came the winter's storm,
As if abashed, its fury checked.
No more above and round the path,
Beneath the wind's tempestuous wrath,
The snowy billows heave and toss;
A sacred calm as he draws nigh
Pervades at once the earth and sky.
His robe was blue, its borders decked
With evergreen and scarlet moss;
His hands upon each other rest,
Due north and south, due east and west;
The open palms together pressed
As if engaged in silent prayer.
He thus had formed with pious care
The holy symbol of the cross.
A lamb doth close beside him go,
Whose whiter fleece rebukes the snow:
These things sufficiently proclaim
His mystic office and his name.
Beside the grave he comes and stands,
Still praying there with folded hands;
And, while he prays, see drawing near
Another shining form appear,
His right hand on his bosom pressed,
As if by bitter grief distressed,
The other pointing to the skies,
And, as he weeps, each radiant tear,
That from his sad and earnest eyes
Falls on the earth, is transformed there
To violets blue and blossoms fair,
That sweetly perfume all the air.[1]
A third one now appears in sight,
Arrayed in royal robes of light,
A "lordly lion" walks in pride.
More glorious far; and at his side
And he who came in glory last
Between the others gently passed,

---

[1] Ebers, the German Egyptologist, informs us that the Egyptians believed the tears of the immortals had this creative power.

And, looking down upon the dead,
With level, open palms outspread,
A holy benediction said.

This done, the *first one, by command,*
The dead god taketh by the hand:
At once through all the body flies
The same warm flush that marks the skies.
The shrunken features, cold and white,
A moment shine with life and light.
A moment only—'tis in vain:
Unconquered Death resumes his reign.

So doth a solitary wave
Leap up amid the lonely night,
And catch a gleam of life and light,
And then sink helpless in its grave.
To raise the god the first thus failed—
The powers of darkness yet prevailed;
So to the *second* he gives place,
Who, like the first one, by command,
The sun-god taketh by the hand,
And, looking downward in his face
With pleading voice and earnest eyes,
On Phoebus calls and bids him rise.
Though at his touch the blood unbound.
With rapid current red and warm
Runs swiftly through the prostrate form,
Yet silent on the frozen ground
The god lies in a trance profound,
Devoid of motion, deaf to sound.

Alas! alas! what doth remain?
Shall death and darkness ever reign,
And night eternal hide the day?
Then said the *third one,* "Let us pray."
And full of faith and strong intent,
His prayer to IH. VAH. upward went.
*"Amen"was* said—*"so mote it be!"*
And then the last one of the three
Arose and stretching forth his hand,

## Chapter 6. Allegory of the Death of the Sun

Calls on the dead, and *gives command*
In IH. VAH.'S name to rise and stand.

Then up rose Phoebus in his pride,
With the "lordly lion" by his side,
And earth and sky with his glory shone
As again he sat on his golden throne.
For the voice of God is nature's law,
And strong was the grip of the lion's paw.

# Appendix to Part Second

**THE LION'S PAW---- ANCIENT EGYPTIAN DRAWING**

Since Part I of this work was written, I find in the "Masonic Newspaper," of March 6, 1880, the above emblematic drawing, concerning which Brother William S. Paterson (thirty-second degree) says:

> This emblem was found in the sarcophagus of one of the great kings of Egypt, entombed in the pyramid erected to his everlasting remembrance. It brings to mind the representation of the king's induction into those greater Mysteries of Osiris, held to be the highest aim of the wise and devout Egyptian.

Brother Peterson also says in the same article that

> the Hebrews were probably instructed in the legend of Osiris, and afterward changed the whole to accord with the wonderful and wise Solomon and his master-architect Hiram[;]

and adds that "the discoveries now going on in Egypt may lead to the key of these mysteries." Brother Patterson makes no attempt to explain the hidden meaning of this ancient Egyptian emblem; but, if the theory advanced in this work is correct, the reader will have no great difficulty in understanding it, for the same astronomical key which unlocks the hidden allegory of the legend of Osiris and of Hiram will also fully explain this ancient emblem, while the fact that this emblem so graphically and perfectly illustrates our astronomical solution of the legend is strong corroborative proof of its correctness.

The emblem may be thus explained: the form that lies dead before the altar is that of Osiris, the personified sun-god, whom the candidate represents in the drama of initiation, lying dead at the winter solstice. The cross upon his breast refers to the great celestial cross, or intersection of the celestial equator by the ecliptic. The figure of the lion grasping the dead sun-god by the hand alludes to the constellation Leo and the summer solstice, at which point the sun is raised to life and glory, as has been just explained in the allegory of the resurrection of the sun, and denotes that the candidate is about to be raised from a symbolical death to life and power by the grip of the lion's paw. This is made clearly manifest from the fact that the lion holds in his other paw the ancient Egyptian symbol of eternal life, or the *Cruz Ansata,* a full description of which and its true meaning are given in Part Third (see page 210). The tablet at the feet of the candidate has inscribed upon it in hieroglyphics the sacred names of *Amen* and of *Mat,* the wife of *Amon Ra,* and probably that of the royal candidate. The figure erect at the altar is that of the Grand Hierophant, attired as Isis, with the vacant throne upon her head, emblematic of the departed sun-God. She has her hand raised in an attitude of command, her hand forming a right angle; her eyes are fixed upon the emblematic lion as she gives the sign of command that the candidate be raised from death and darkness to light and life.

The objects on the altar are two of those peculiar-shaped glass jars, with pointed bases, in which wine was kept (See Wilkinson's "Egyptians of the Time of the Pharaohs" page 86, woodcut 62), and which, the same author says, "always had their place on the altar of the gods" (page 13). The emblem placed between the votive jars of wine is more obscure. It may be the *thyrsus,* but is more probably a floral offering. (See "Ancient Egyptians," vol. i, woodcut 260, No. 5.) There can be no doubt but that the whole device is a symbolical picture of the initiation of some important person into the Mysteries, not of Osiris, however, as Brother Paterson thinks, but of Isis, who, represented by the Grand Hierophant, stands behind the altar, giving the command to raise from death Osiris, who lies before it. This ancient Egyptian drawing is a strong and startling testimony of the entire correctness of the astronomical solution of the legend of Osiris and that of Hiram, as given in the foregoing pages. It is indeed, almost impossible to make an emblematic drawing which would be in more perfect harmony with it.

### The Judgment of the Dead

As the judgment of the dead, or Judgment of *Amenti,* formed a part of the Mysteries of Isis, it should be properly mentioned in that connection. Although this ceremony was part of the Mysteries, yet it was well known to all, as it was founded upon the peculiar funeral rites of the Egyptians. From this judgment in this world no Egyptian was exempt, no matter how exalted his position; and upon this trial depended the right to an honorable burial. All whom the deceased person had wronged, and all who knew of his evil deeds, were permitted to testify over his dead body, while his friends and kindred loudly proclaimed his virtues. The decision followed the weight of the evidence; and even a king who had led a bad and wicked life might be excluded from burial in his own sepulchre. And the "assessors" at the funeral were allowed to pronounce a condemnation, which all agreed would also be received in a future state. This trial of the dead in this world was typical of the judgment of Amenti, where Osiris presided in the invisible world, and which the devout Egyptian believed took place there at the same time.

From this peculiar custom of the Egyptians arose a part of the ceremonies of initiation into the Mysteries of Isis; for, as in initiation, the candidate died symbolically, so also he underwent the Judgment of the dead, to ascertain if he was worthy to receive the higher and more important secrets, by being raised and brought to light, typical of the admission of the good into the "mansions of the blessed." The last judgment is one of the principal subjects found depicted upon the walls of tombs and in the "Book of the Dead," sometimes referring to the actual trial, at others to its representations as enacted in the Mysteries. This judgment of the dead was peculiar to the national customs and funeral rites of the Egyptians, and does not appear to have prevailed in other countries. It was therefore naturally discontinued as a part of the Mysteries when they were introduced into other countries other names. The Greeks, however, introduced it into their mythology—the Greek Themis being derived from the Egyptian *Themei,* or goddess of Justice; while Minos and *Rhad-amanthus,* the Grecian judges of the dead in Hades, show their connection with Amenti, the Egyptian Hades, or region of darkness. The transport of the body over the sacred lake in the *baris,* or boat, in the funeral procession of the Egyptians, in like manner, gave rise to the Acherusian lake, the ferryboat of Charon, and the passage of the Styx, in the Grecian mythology. There is nothing in the ancient masonic degrees in the least analogous to the Judgment of Amenti, that portion of the Mysteries of Isis not having been adopted into the Mysteries as celebrated in other lands and at later age.

The following representation of the scene, taken from the "Book of the Dead," will, however be interesting to all readers, and members of the fraternity will not fail to recognize in it certain masonic features which we may not particularize. The figure seated on the throne of Osiris, or judge of the dead; he

THE JUDGMENT OF AMENTI

holds the flail and crook, emblems of majesty and dominion. The deeds of the deceased, or of the candidate, typified by a vase containing his heart, are being weighed in the scales of justice by Anubis and Horus against an ostrich-feather, emblem of truth, in the opposite scale. The ostrich-feather, as the emblem of truth, is thus depicted in the hieroglyphics: Thoth (Hermes, Mercury, or the Divine Intellect) presents the result to Osiris. Close by is Cerberus, guardian of the gates. Below the candidate is such attended by the goddesses of Truth and

EMBLEM OF TRUTH

Justice; the goddess of Truth holds in her hand the emblem of eternal life, and both wear upon their hands the emblem of truth. Close to Osiris is seen the *thyrsus* bound with a fillet, to which the spotted skin of a leopard is suspended. It is the same that the high-priest, clad in the leopard-skin dress, carries in the processions, and which gave rise to the *nebris* and *thyrsus* of Bacchus, to whom Osiris corresponds in Greek

102

mythology (Wilkinson). The lotus-flower, the emblem of a new birth, is represented just before the *thyrsus.* If on being tried, the candidate is rejected, having been "weighed and found wanting," Osiris inclines his scepter in token of condemnation. If, on the contrary, when the sum of his deeds has been recorded, his virtues so far preponderate as to entitle him to admission, Horus, taking in his hand the tablet of Thoth, introduces him to the presence of Osiris. In the initiation, those who represented Thoth, Anubis, and Horus wore symbolical masks, as represented in the drawing. (See Kendrick, Wilkinson, and also Arnold's "Philosophical History of Secret Societies," from which last work the above drawing is taken.)

# Part Third

## Chapter 7

# *Astronomical Explanations*

*of the Emblems, Symbols, and Legends of the Mysterles, Both Anclent and Modern, and the lost Meanlng of Many of Them Restored.*

**HAVING** EXPLAINED the solar allegory which is embodied in the legend of Hiram, as well as the solar symbolism attaching to the officers of the lodge, their several "stations" and duties, and the "lights, covering, and supports of the lodge," it now remains to consider the various emblems and other legends of freemasonry. If it can be shown that all of them (which are at all ancient) have also a solar and astronomical allusion, in perfect harmony with each other and with the main central legend which they are intended to illustrate, the fact that the whole system is founded on an astronomical allegory will be irresistibly forced upon us. The various emblems, symbols, and legends dependent on that of Hiram, and intended to illustrate it, will therefore next claim attention. In this examination the same method of question and answer will be pursued, as being best adapted to the object in view:

Q. Have all the ancient signs, symbols, emblems, and legends of the Mysteries, both ancient and modern, an astronomical allusion?

A. They have. As the whole system has an ancient astronomical foundation, it could not be otherwise.

Q. Has the astronomical allusion of many of the emblems, symbols and legends been lost?

A, It has; as to some, entirely, and as to others, in part. The allusion is, however, so perfect in most cases, that it may be restored by the use of the key already furnished to the main central allegory, to which they are all harmoniously related.

### *The Seven Stars*

Q. To what does the masonic emblem of the seven stars allude?

A. To the *Pleiades,* or seven stars in *Taurus.* These stars are called by the Romans *Vergilioe,* or Virgins of Spring. The constellation *Taurus* was anciently at the vernal equinox, and the year formerly then began. Thus Virgil, referring to a remoter age, in the "Georgics," Book I says:

> "Candidus auratis aperit cum
> Cornibus annum taurus."

> "When the bright bull with
> gilded horns opens the year."

Job speaks of the Pleiades, also, as exerting "a sweet influence," expressive of the balmy air of spring which accompanies the approach of the sun to the constellation Taurus and the "seven stars." This masonic emblem, therefore, has a direct allusion to the vernal equinox, and thus becomes a beautiful symbol of immortality, reminding us, also, of that starry home beyond the grave to which the soul of man aspires. It was for these reasons that, of all the "hosts of heaven," the Pleiades were selected as an emblem by our ancient brethren.

## *The Ladder of Seven Rounds*

Q. What is the true meaning of this ancient emblem?

A. The ladder of seven rounds, says the London "Freemason's Magazine," has been a symbol in many ages and countries.

> Among the ancients every round was considered to be represented by a metal increasing in purity, from the lowest to the highest, and these were again characterized by the names of the seven planets as follows: The first round is the lowest; therefore they will read from the bottom to the top:
>
> 7.  Gold — the Sun;
> 6.  Silver — the Moon;
> 5.  Iron — Mars;
> 4.  Tin—Jupiter;
> 3.  Quicksilver — Mercury;
> 2.  Copper — Venus;
> 1.  Lead — Saturn.

This planetary signification given to the seven rounds of the ladder, as stated by the writer of the above, is in perfect harmony with the religious ideas of the ancients who worshipped the sun and planets, and the several allegorical legends which they founded upon the facts of astronomical science.

Near the site of ancient Babylon are the ruins of the great Temple of the Seven Spheres, which for a long time was thought to identical with the great Temple of *Belus,* described by Herodotus, situated in Babylon, and which it closely, though not exactly resembles. The builder of this temple is unknown, and the date of its original structure is also uncertain. It was, however, restored and carefully renovated by Nebuchadnezzar, whose name is still legible on the bricks and cylinders deposited at the angles. The account which the royal restorer gives of his work has been likewise found on the inscriptions among the ruins. The following particulars as to this great temple, which is a type of the plan and character of all the Babylonian sacred buildings, is taken from Rawlinson's Appendix to Book III of Herodotus. The ruins were carefully and completely explored by Sir H. Rawlinson himself but a few years ago. Like the great Temple of Belus at Babylon, as described by Herodotus, the Temple of the Seven Spheres was a building of *seven receding stages.* At the top of the seventh stage was placed the *ark,* or tabernacle, which seems to have been fifteen feet high. The ornamentation of the building was almost solely by color.

> The seven stages were colored so as to represent the *seven planetary spheres,* according to the tints regarded by the Sabaens as appropriate to the seven luminaries, the basement being black, the hue assigned to Saturn; the next an orange, the hue of Jupiter; the third a bright red, the hue of Mars; the fourth a golden hue of the Sun; the fifth a pale yellow, the hue of Venus; the sixth dark blue, the hue of Mercury; the seventh silver, the hue of the Moon.

From the fact that the seven stages by which the summit of the temple was reached were thus dedicated to the seven planets, it is evident that the symbolism of the seven steps of the ladder, and the seven ascending stages of the temple is the same. The *order* in which the planets are arranged is, however, not exactly the same as that of the steps of the ladder as given by the "Freemasons' Magazine" of London. The latter seems to be founded mostly on the supposed order of the metals as to purity. That the order of the planets, as applied to the seven stages of the temple, is the most correct according to the ancient symbolism of the Babylonians and other Oriental nations, can not be doubted, for the ruins of the temple itself place that beyond question. It is also equally evident, from the description of Herodotus, that the symbolism of the seven stages of the Temple of the Seven Spheres is the same as that of the great Temple of Belus itself at Babylon. One of the principal emblems of the ancient Mysteries, both of Persia and India, was a ladder of seven rounds or steps, and it may be traced back to the very builders of those temples. In attempting to ascertain the true meaning of this emblem, we must not forget that the Babylonians and the Sabaens were worshippers of the planets. It is also equally important to remember that they were adepts in astronomy, and believed in and practiced astrology. This is evident from sacred history. We read in Daniel ii that Nebuchadnezzar (the same who rebuilt or restored this very Temple of the Seven Spheres) was troubled by a dream, which he commanded "the magicians, *the astrologers,* and the sorcerers, *and the Chaldaens,* to interpret for him." (See also v. 7, and many other passages of Scripture.)

> The Chaldeans were a branch of the great Hamite race of *Ak Kad,* which inhabited Babylonia from the earliest times. With them originated the art of writing, the building of cities, the institution of religious systems, the cultivation of all science, but that of *astronomy* in particular. (H. C. Rawlinson)

The sciences of astronomy and astrology will, therefore, no doubt furnish the key to the symbolism of not only the seven stage of the temple, but the seven rounds of the ladder also. In truth, the reference in both to the seven planets points us earnestly in that direction.

The sun on the 21st of December is at his lowest point of declination below the equator, and the days are the darkest and the nights the longest, while all nature lies dead, locked in the arms of winter. On the 21st of March the sun reaches the vernal equinox. Spring begins, and nature revives from the death of winter. On the 21st of June the sun reaches the summer solstice, when the days are the longest, and the sun seems for the first time to have regained all his former power and glory. Now, it will be observed, by looking at any celestial globe, that the progress of the sun from its *lowest to* its *highest* declination is divided into seven equal parts by the seven signs of the zodiac, through which he passes, or in which he is, while mounting upward from the winter to the summer solstice. The sun, starting in *Capricornus,* passes successively through *Aquarius, Pisces, Aries, Taurus,* and *Gemini,* until he reaches the summer solstice, or summit of the zodiacal arch, on the 21st of June. If the reader will take the trouble to trace this ascending path of the sun along the ecliptic on a celestial globe, its symbolical significance will be impressively illustrated. It must, however, be remembered that the winter and summer solstice were anciently in *Aquarius* and *Leo,* and not in *Capricornus* and *Cancer,* as they now are, owing to the precession of the equinoxes.

The Hindu astronomer, Varaha, says, "Certainly the southern solstice was once in the middle of *Asleha (Leo),* and the northern in *Dhanishta (Aquarius)."* Modern astronomers all declare the same thing. A study of the various astronomical myths of antiquity shows that the most of them originated when the summer solstice was either in Leo or between Leo and Cancer. In the days, therefore, when planetary worship had its rise, the sun, in his passage from the winter to the summer solstice, started in *Aquarius* and ascended successively through the signs *Pisces, Aries, Taurus, Gemini,* and *Cancer,* 30° each, and entered *Leo* at the summit of the zodiacal arch on the 21st of June. These seven signs are therefore symbolical of seven ascending stages or steps, and, according to the science of astrology, these seven signs, following each other in this exact order, are the houses of the seven planets (which they rule and signify) in exactly this order: Saturn, Jupiter, Mars, Venus, Mercury, the Moon, and the Sun.

But by a strange correspondence this is the exact order in which the planets are arranged as ruling the seven ascending stages of the Temple of the Seven Spheres, with but one exception. The moon, whose house is *Cancer,* and which sign she rules (according to astrology), is at the top of the seven stages, while the sun is placed in the center, between Mars and Venus, who rule the vernal signs Aries and Taurus. This is, however, in perfect harmony with the ancient allegory above given, for anciently the vernal equinox was between *Aries* and *Taurus,* the summer solstice being between *Leo* and *Cancer.* The base of the temple, therefore, symbolized the winter solstice—the appropriate color of which was black, and its significator *Saturn* or Time, which destroys all things. It referred to the sun at his lowest point of declination, and when Nature is desolate and dead.

The central stage, ruled by the sign *Aries* and *Taurus,* between which the sun was emblematically represented by his color, was typical of that luminary raised to life again at the vernal equinox, when the sun entered those signs in the spring. The seventh stage, or summit of the temple, was in like manner typical of the summer solstice, anciently between *Cancer* and *Leo. Cancer is* ruled by the moon, and *Leo* is the sole house of the sun (according to the teaching of astrology).

The top and last stage was therefore represented to be of the color of the moon, denoting that the sun was now approaching the highest point of his journey, and was about to be exalted to the summit of the zodiacal arch. The colors, as given by Herodotus, are also in exact harmony with the science of astrology, and so also is the rule of the seven metals by their respective planets, as given by the "Freemasons' Magazine" (see Ptolemy, Placidus, Lilly, and Zadkiel's "Grammar of Astrology," for the teachings of astrology on these points; also, as to the houses of the planets and their rule). We should be pleased to follow this subject still further, but enough has been said to show the close connection between the *seven ascending stages* of the great Temple of Belus and the Temple of the Seven Spheres at Babylon, with the emblem of the *ladder of seven steps* as exhibited in the Persian Mysteries, and, indeed, all of the Oriental Mysteries. Nor can there be much doubt of the fact that our masonic emblem was adopted from these ancient sources, while it is equally certain that the explanation which refers it to "the ladder Jacob saw in his vision," although beautiful, is the invention of Preston, Cross, or some
112

other recent writer, who had no idea of its true meaning or ancient origin.

## *The Masonic Ladder of Three Rounds*

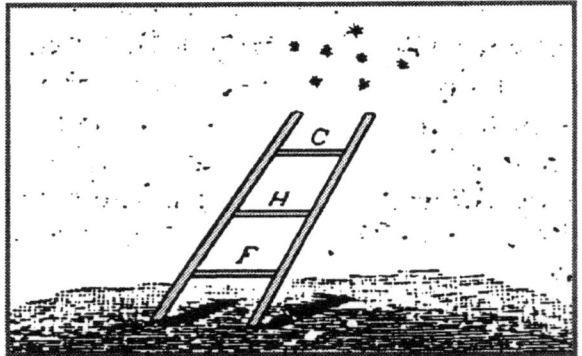

Q. What is the signification of the ladder of three rounds, and why is it represented as leading up to the "seven stars," or Pleiades?

A. This emblem is clearly but a modification of the ladder of the Mysteries, consisting, as we have seen, of seven rounds— and is of the same general astronomical meaning. The sun, when ascending from the winter solstice to the vernal equinox, the constellation Taurus ($\Upsilon$), and the Pleiades, or seven stars, situated therein, passes successively through three signs of the zodiac, to wit, Aquarius ($\approx$), Pisces ($\mathcal{H}$), and Aries ($\Upsilon$).

**THE MASONIC LADDER OF THREE ROUNDS**

These three signs are therefore emblematically represented by a ladder of three principal rounds, by means of which the sun climbs up from the point of his lowest southern declination to the vernal equinox and the "seven stars" in Taurus. The foregoing is the emblem of the masonic ladder as generally represented (see Monitors).

The diagram following will show how perfectly the explanation of its meaning, as given above, agrees with all the facts of astronomy, and how significant and beautiful the emblem is when thus considered.

## *The Zodiacal Ladder*

The emblematic meaning now attached to the masonic ladder, which refers it to the one "Jacob saw in his vision," is neither lost nor sacrificed, even if we admit the probable origin of the emblem in that of the ancient mysteries. Its symbolism is, however, thus made more extended and impressive, so that we gain rather than lose by so referring it.

**THE ZODIACAL LADDER**

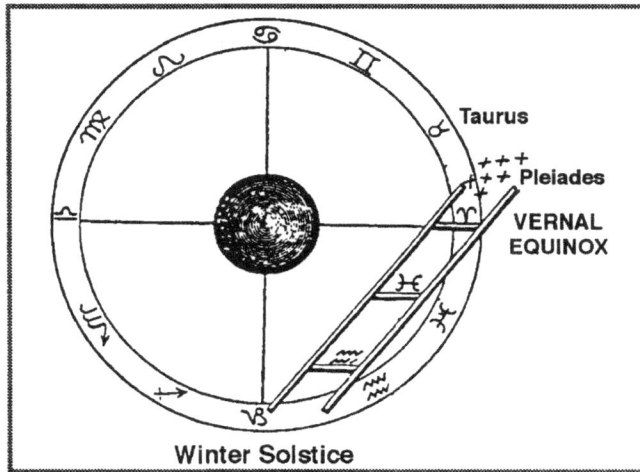

Taurus

Pleiades

VERNAL EQUINOX

Winter Solstice

The initiation into all the ancient mysteries, it will be remembered, was a *drama* founded upon the astronomical allegory of the death and resurrection of the sun, and was intended to, and did, impress upon the mind of the candidate, in the strongest manner possible, the two great doctrines of the unity of God and the immortality of man.

These are today the two great fundamental principles of Freemasonry, and are illustrated and taught in a similar manner in the ritual of the third degree.

The solar allegory and emblems of the ancient mysteries have, however, a twofold meaning:

1. Being founded, as before stated, on the passage of the sun among the twelve constellations of the zodiac—his overthrow by the three autumnal months, his return to life at the
114

vernal equinox, and his exaltation at the summer solstice—
they therefore taught and illustrated all the leading principles
of astronomy, and thus had an important *scientific* value to the
initiated.

2. By personifying the sun, and requiring the candidate to
represent him, the whole solar phenomena were exhibited in
an allegorical manner, and became symbolical of the unity of
God and the immortality of the soul. The ladder of the Mysteries, being but an emblem intended to illustrate the main solar
allegory, had the same two-fold symbolism.

When fully explained to the initiated, it fixed upon the
mind certain great facts in astronomical science. It taught the
order and position of the signs of the zodiac; the ascent of the
sun from the point of his lowest declination below the equator
to that of his highest above it, by seven equal graduated steps.
It also taught the duration and order of the seasons, the length
of the solar year, and many other particulars of the greatest
importance to agriculture, as well as to science and art
generally.

The emblem, viewed in an allegorical sense, also taught,
by solar analogy, the unity of God and the life everlasting. The
ladder in this sense was the emblem of the ascent into heaven
from the lower hemisphere—the underworld of darkness,
winter, and death. This mystic ladder leads to the "seven
stars," or *Pleiades,* shining in the constellation *Taurus,* at the
golden gates of spring. It mounted still onward and upward, to
the summit of the Royal Arch of heaven, thus emblematically
teaching us that by the ladder of virtue the soul of man will at
last pierce the "cloudy canopy," and mount to the highest circle of "the starry-decked heavens," to dwell for ever triumphant over death and the grave.

It will thus be seen that our masonic emblem loses none of
its significance by its probable origin in the astronomical symbolism of the ancient mysteries, but, on the contrary, has given

it a much more extended and beautiful signification, being clothed with a scientific as well as a moral meaning.

## Faith, Hope, and Charity

Q. Why may the three principal rounds of this ladder be also said to emblematically represent "faith, hope, and charity?"

A. When the sun has reached his lowest southern declination, and begins to ascend toward the vernal equinox, we have nothing *but faith* in the goodness of God and the immutability of the laws of nature to sustain our belief that the sun will once more "unlock the golden gates of spring"; but, when the sun enters Pisces (♓) and ascends the *second* round of the ladder, *hope* is added to our faith, for the sun is seen already to have climbed up two thirds of the distance required to reach the vernal equinox; and when, at last, on the 21st of March, he mounts the *third* round of the ladder and enters Aries (♈), the "sweet influences of the Pleiades" are once more felt, while beneath the warm rays of the vernal sun the snows dissolve, and the earth begins again "to put on her beautiful attire." "For lo! the winter is past, and the flowers appear on the earth, and the time of the singing of birds is come, and the voice of the turtle is heard in the land." The third and last round of the zodiacal ladder is therefore emblematic of *charity,* or that divine love and benevolence which each year cause the springtime to come in due season. So ought we all to have faith in God, hope in a blessed immortality (emblematically represented by the vernal equinox), and charity to all mankind.

### The Three Steps

The three steps delineated on the master's carpet have an obvious reference to the three steps, or degrees, by which the initiated becomes a master mason. They are, however, capable of an astronomical explanation also, and may be said to allude to the three signs, *Taurus, Gemini,* and *Cancer* (emblematic of three steps), by means of which the sun (having already

116

reached the vernal equinox by means of the zodiacal ladder) ascends to the summit of the Royal Arch at the summer solstice, which point is, as already explained, emblematic of the master's degree.

## The Winding Steps

Q. According to the legend of the "middle chamber" of the fellow-craft's degree, the workmen were paid their wages in the middle chamber of King Solomon's temple, which was approached by a certain flight of *"winding steps."* This staircase is said to have consisted of "three, five, and seven steps" (according to our lecture), and was reached by entering in at the front door of the temple, passing between the pillars of the porch. (See Mackey's "Symbolism," Chapter XXVI.) What is the astronomical import and real meaning of this legend?

A. The only allusion to these "winding stairs" in the Bible is found in the sixth chapter of 1 Kings. In the fifth verse we are informed that King Solomon "built chambers *round about against the walls of the house"* The sixth verse continues as follows:

> The nethermost chamber was five cubits broad, and the *middle chamber was* six cubits broad, and the third was seven cubits broad, for *without in the walls of the house* he made narrow rests round about, that the *beams should not be fastened in the walls of the house.*

The eighth verse informs us that the

> *door for the middle chamber* was on the *right side* [Hebrew, "shoulder"] of the house, and they went up with winding stairs into the middle chamber, and out of the midst of the middle into the third.

The only information which Josephus gives may be found in Chapter III, Book VIII, of his "Antiquities," and is as follows:

He [Solomon] also built about the temple thirty small rooms, which might include [i.e., surround] the whole temple by their closeness one to another, and by their number and *outward position round it.* He also made passages through them, that they might come into one through another. Every one of these rooms had five cubits in breadth, and the same in length, but in height twenty. Above these were other rooms, and others above them, equal both in their measures and numbers, so that these reached to a height equal to the *lower* part of the house, for the upper part had no buildings *about it.* The roof that was over the house was of cedar: and, truly, every one of these rooms *had a roof of its own that was not connected with the other rooms,* but for the *other parts* there was a covered roof common to them all.... The king had also a fine contrivance for an ascent to the upper room over the temple, and that was by *steps cut in the thickness of the wall,* for it had no large door on the east end, as the lower house had, but the entrances were *by the sides through very small doors.*

The above extracts comprise all the information which reliable history, either sacred or profane, furnishes in regard to the *"middle chamber"* and the *"winding stairs"* by which it was reached. It is evident, both from the Bible and from Josephus, that the "middle chamber" and the "winding stairs" by which it was reached. It is evident, both from the Bible and from Josephus, that the "middle chamber" was no part of the temple proper; nor, indeed, was it permitted to be fastened to the sacred walls. (See 1 Kings 6, 5, just quoted.) All the chambers were built around the outside of the walls, and were reached from the side, so that in going up to the "middle chamber" a person not only did not pass between the pillars of the porch, but did not enter in or pass through any portion whatever of the temple itself. The steps, according to Josephus, were "cut in the thickness of the wall outside." In view of these authorities, although he does not quote them, Dr. Mackey may well say

that the historical facts and the architectural details alike
forbid us for a moment to suppose that the legend [of
the winding stairs], as it is rehearsed in the second
degree of masonry, is anything more than a magnificent
*philosophical myth.*          (Symbolism," Chapter XXVI)

But if it is a "philosophical myth" it must have a symbolical
meaning, and be emblematic in its character. The very essence
of symbolical teaching consists of the method of selecting
some *fact* or some *real object in* nature, art, or science, and by
investing it with an emblematic significance through comparison, thus teaching and illustrating some moral or political doctrine. The anchor is thus made an emblem and illustration of
hope, the beehive of industry, the scythe of time or death. A
real anchor, beehive, or scythe is, however, required as a
foundation for this allegorical teaching. If, therefore, the "legend of the winding stairs" is a "philosophical myth," either the
actual or the emblematic stairs must have a real existence
somewhere, or they could not have been selected or used for
the purpose of conveying a philosophical, symbolical, or allegorical lesson. The "winding steps," as described in the
masonic legend, did not exist in the temple of King Solomon,
as we has shown, not only by Josephus, but the bible itself.
We must, therefore, look elsewhere for them. Now as all the
other leading emblems of masonry have an astronomical
orgin, it is but reasonable to suppose that these very same
"winding steps," leading to the place where the wages of the
craft are paid, will be found in that other "temple not made
with hands, eternal in the heavens." As they are not to be
found in the actual temple, let us look for them in the *emblematic* one.

But, before doing so, it will be necessary to determine
more exactly the proper number of these emblematic steps,
for their stated number seems to have varied at different periods and according to different versions of the legend.
Dr. Oliver mentions an old "tracing-board," published in 1745,

in which the steps are *semicircular,* and are but *seven* in number. Dr. Mackey says, on page 221 of his "Symbolism," that

> tracing boards of the last century have been found in which only *five* steps are delineated, and others in which they amount to seven. The Prestonian lectures used in England in the beginning of this century gave the whole number as *thirty-six,* divided into series of one, three, five, seven, nine, and eleven.... The Hemming lectures, adopted by the union of the two grand lodges of England, struck out the eleven.... In the United States the number was still further reduced to *fifteen,* divided into three series of three, five, and seven.

It thus appears that there has been considerable confusion as to the correct number of these symbolical steps. The most ancient versions of the legend make the number either five or seven. Now it is a very safe rule to adopt as to all traditions, including those of masonry, that the older the version the more correct it probably is, for the further back we trace any legend the nearer we will approach the time of its origin, and, consequently, its primitive and uncorrupted form. Applying this rule to the case under consideration, we may safely conclude that the proper number of steps in those "winding stairs" is either five or seven. If, however, we succeed in finding the steps themselves properly located in the emblematic temple, and leading to the very place where the craft receive their wages, we shall be able to determine their exact number by actual count.

The building of the temple, represented emblematically by the Royal Arch of heaven, was commenced in the spring and finished in the autumn. It was, therefore, said to be seven years in building, as has been previously explained. The spring signs, during which the plowing and planting are done, are typical of the E. A. degree; the summer months, when the growing grain requires constant care for its protection, of the F. C. degree; and the season in which the harvests are gathered and stored away, of the M. M. degree, and those skilled work-men who wrought at the completion of the temple.

During the progress of the sun from the vernal equinox to the summer solstice, the husbandman is engaged in preparing the soil and sowing his seeds; during the passage of the sun from the summer solstice to the autumnal equinox, he is employed in protecting his maturing crops. In July and August the *corn* ripens and is harvested, and in the autumn the *oil* and w*ine* also reward him for his labors.

The wages of the faithful craftsmen, we are told, are *"corn, oil, and wine."* The *seven* signs of the zodiac, from the vernal equinox to the first point of *Scorpio,* "winding" in a glittering curve about the heavens, may in a like manner be said to be emblematic of seven winding steps, leading to the place where—

### *Corn, Oil, and Wine*

—are brought forth to reward the labors of the husband-man. The sun arrives at Aries on the 21st of March, and reaches Scorpio about the 21st of October, passing succes-sively through ♉, ♊, ♋, ♌, ♍, and ♎. The number of these emblematic steps is therefore *seven,* thus corresponding with the more ancient versions of the fellow-craft legend; and it will also be observed that they are really *semicircular in* form. This perfectly harmonizes with the "seven semicircular steps" of the ancient "tracing-board" mentioned by Dr. Oliver. It is also wor-thy of notice that, just as that part of the year embraced within these seven signs may be divided into three periods—1. That of plowing and planting; 2. That of growing and maturing; and, 3. That of harvesting and storing—so these emblematic steps may also be divided into three groups, which find an appropriate expression in the numbers 3, 5, and 7. The first *three* signs, Aries, Taurus, and Gemini, denote the season of plowing and planting. The next, *two,* Cancer and Leo, making *five* from the vernal equinox, denote the period during which the crops ripen and mature; and the last *two,* Virgo and Libra, making seven in all, rule the harvest-season—and the storing away of the *corn, oil,* and *wine,* with which the solar bounty has rewarded the labors of the faithful husbandman.

The American division of the steps into three groups, expressive of the numbers 3, 5, and 7, is therefore correct, but the *total* number of steps is seven, and not fifteen. It is easy to see how this latter error, as to the mystic import of the numbers 3, 5 and 7 was made, in consequence of the true nature of the symbolism of the seven steps being lost.

The legend of the "winding stairs" informs us that they conducted between the two pillars of the porch. Dr. Oliver, in his "Landmarks" (note 19 to Lecture XVI), says that "the equinoctial points are called *pillars,* because the great semicircle, or upper hemisphere, seems to rest upon them." If this symbolism be correct, then, the "winding stairs" do, in fact, lead past and between these celestial pillars, in perfect harmony with the allegory of the legend. Thus explained, the legend of the "winding stairs," leading to the place where "corn, oil, and wine" are delivered as a reward to the faithful laborer in the vineyard, is a most beautiful and significant astronomical allegory. Like all the other astronomical allegories and symbols of Freemasonry, it not only (when properly understood) reveals important and valuable *scientific* facts respecting the movements of the heavenly bodies, but at one and the same time inculcates, in a sublime and impressive manner, great *moral* truths. It teaches us, among other things, that industry not only deserves but receives its due reward. It also displays the benevolence of the Great Creator, who causes the earth to bring forth her fruits in due season:

> He watereth the hills from above;
>> The earth is filled with the fruit of his works;
> He bringeth forth the grass for the cattle,
>> And the green herb for the service of man:
> That he may bring forth fruit out of the earth;
>> And wine, that maketh glad the heart of man;

> And oil, to make him a cheerful countenance;
> And bread, to strengthen man's heart.
>
> *Psalm 104:15-15*

It reminds us also of the covenant which God made with Noah in the olden time: "That he would no more curse the ground for man's sake; but that while the earth remained seedtime and harvest should not cease" (Genesis 8:21-22). These and many other important lessens are taught by the astro-masonic symbol of the "winding stairs"; and those lessens are made still more impressive from the fact that the Archetype of these "winding stairs" is not to be found in any transitory, earthly mansion, but far above, set in the eternal majesty of the starry firmament.

### The Blazing Star

Q. To what does the masonic emblem of the Blazing Star
   allude?

**THE
BLAZING
STAR**

A. To the sun the midst of heaven, as a symbol of Deity. Even Dr. Oliver, who has no sympathy with the astronomical theory of the origin of Freemasonry, says:

> The *"Blazing Star"* must not be considered merely as the creature which heralded the appearance of T. G. A. O. T. U., but the expressive symbol of that *Great Being himself,* who is described by the magnificent appellations of the Day-Spring, or *Rising Sun,* the Morning Star. This, then, is the supernal reference of the *Blazing Star of Masonry,* attached to a science which, like the religion it embodies, is universal and seasons, and to every

people that ever did or ever will exist on our ephemeral globe of earth.

ANUBIS

Other writers identify the Blazing Star with Sirius, the most just before the sun, each year, gave the ancient Egyptians warning of the approaching inundation of the Nile; hence they compared it to a faithful dog, whose bark gives warning of approaching danger, and named it *Sothis, Anubis,* and *Thotes,* the barker, or monitor. This brilliant and beautiful-star thus early became known "dog-star". The Egyptians deified it under the name of Anubis, and this god was emblematically repre-sented by the figure of a man with a head of a dog.

Both these explanations show the masonic Blazing Star to be an astronomical emblem. The latter is probably the more correct, as it appertains to the Egyptian Mysteries.

### The Rite of Circumambulation

Q. To what this masonic rite allude?

A. The word "circumambulation" is derived from two Latin words *(circum,* around, and *ambulare,* to walk), and therefore means to walk around, that is, around the altar, or some sacred shrine. The rite of circumambulation formed a leading part of the ceremonies of the Mysteries, and of solar worship in all countries. This rite had a direct solar allusion, as it was always performed from right to left in imitation of the apparent course of the sun from east to west by way of the south. In the Mysteries of India the candidate went thus about the altar three times, and, whenever he arrived in the south, was taught to exclaim, "I copy the example of the sun, and follow his benevolent path!" This sacred march was generally, in all the Mysteries, accompanied by the singing or chanting of an ode or hymn to the sun-god. Among the Druids it partook of the nature of a mystic dance. The candidate, in performing the rite of circumambulation, it will be seen, represented the sun, or rather the personified sun, or sun-god, which he continued to do through the entire ceremony, from the moment of his introduction up to his symbolical death—*Euresis* and raising or restoration to life. Dr. Mackey says, in his "Symbolism of Freemasonry," Chapter XXI, that "the masonic rite of circumambulation strictly agrees with the ancient one," and that, as

> the circumambulation is made around the lodge just as the sun was supposed to move around the earth, we are brought back to the original symbolism" of the sun's apparent course about the earth.

The direct derivation of this masonic rite from the solar mysteries of the ancients is too plain to be for a moment denied; and it is absurd to suppose that any such rite could have been invented by the traveling operative masonic associations of the middle ages. And this absurdity will attach to the whole ceremony of which this rite is but a part (in fact, almost the initial step), for the same solar significance characterizes the whole ritual, all parts of which are in perfect harmony with the symbolism admitted to be connected with the rite of circumambulation. If Freemasonry, therefore, originated with the traveling masons of the middle ages, they must have borrowed these solar ceremonies from some far more ancient source, or association, to which those who instituted modern Freemasonry belonged. Had they invented a ritual, its ceremonies would never have had any such solar significance or symbolism: a symbolism which has no harmony or correspondence with the rules and principles of architecture. On the other hand, if, for peculiar reasons, these operative masons and architects really became the last and sole custodians of the rites and ceremonies of the ancient mysteries, we can quite easily see how they have been handed down to us in a more or less corrupted form by them.

What those circumstances were that thus connected the architects of the middle ages with the ancient mysteries will be treated of more at length in subsequent pages; and the link which thus united the temple builders of Egypt, Greece, and Rome with the cathedral-builders of Europe under the reign of Christianity will be pointed out.

### The Square

Q. Whence was the square, as a masonic emblem, derived?

A. It is a general impression, among masons and others, that the square, or right angle, as an emblem, was derived wholly from operative masonry, and is but one of the working tools of a mechanical art adopted as an emblem by speculative masons. This idea is countenanced by Cross in his "American Chart," who says, "The square is an instrument made use of by operative masons to square their work," and then proceeds to moralize upon it. This idea has also found its way into all the monitors. The square, or right angle, as an emblem is, however, geometrical and not mechanical in its origin, and dates back to the ancient Egyptians, in whose solemn processions the Stolistes carried the cubit of justice, by which perpendiculars, right angles, and squares might be laid out, its form being

**CUBIT OF
JUSTICE**

that of one arm of a square, with the inner end cut to an angle of 45°, or one half of a right angle. The square was in Egypt an emblem of justice, because being a right angle it deviated in no respect from a *true* horizontal joined to a *perfect* perpendicular. The close analogy between justice and that which is perfectly upright is so obvious, in fact, as to have become universal. The terms "an upright man" and a "just man" are in nearly all languages synonymous, hence the Scriptural phrases: "The way of the *just* is *uprightness:* thou, most upright, dost weigh the path of the just" (Isa. 26:7); "He that walketh uprightly" (Psalm 15:2); and the admonition "to walk uprightly before God and man." Besides this, the square was used in Egypt to redetermine the boundaries of each man's possessions when, as frequently happened, the landmarks were swept away by the inundation of the Nile, thus recovering to every man his just rights. The Egyptian land-measure itself was an aroura, or a *square,* containing one hundred cubits. (Wilkinson's "Egypt.")

The square, or right angle, represents 90°, or the fourth part of a circle, and has a direct allusion to division of the ecliptic and celestial equator into four equal parts, indicative of the solstitial and equinoctial points, and the division of the year into four seasons. By it we are also enabled to divide the circle of the horizon into quadrants, and by the aid of the sun in the south to correctly mark out the four cardinal points of the compass. In not only geometry, but astronomy also, the use of the right angle is indispensable; and, as its use was thus connected not only with the loftiest problems of science, but with religion also, it soon became universally adopted by the ancients as a sacred emblem, not only of justice, but of rectitude of conduct. As every perpendicular forms a right angle with its base, and is a straight line, so the primitive roots of the words right and wrong mean straight and crooked, or oblique.

## *Masonic Festivals*

Q. What was the origin of the two great masonic festivals, held formerly on the 24th of June and 27th of December in each year?

A. The celebration of those days was purely astronomical in its origin, and refers to the summer and winter solstice. The summer solstice, on the 21st of June, was celebrated as a great solar festival by the ancients, because at that time the sun was exalted to the summit of the zodiacal arch, and attained his greatest power and glory. The arrival of the sun at the winter solstice in December and the commencement of his return north toward the vernal equinox was also celebrated in an appropriate manner. The sun was then considered (according to another allegory) to be new-born, and the moment of his emerging from the constellation which marked his lowest declination was celebrated as the hour of his nativity. At this period, says Macrobius, "the day being the shortest, the god seems to be but a feeble child." After that, he begins to grow, as some say, nourished by a goat, alluding to the constellation Capricorn, and the days begin to lengthen. The great festival of the new birth of the sun was therefore celebrated at this period. These festivals, originally observed on the days of the summer and winter solstices, came in time, owing to the variation of the calendar (as before explained), to be celebrated on the 24th of June and 27th of December instead of the 21st of those months. Modern masons, however, dedicated these days respectively to St. John the Baptist and St. John the Evangelist, who, it is alleged, were born the one at the summer and the other at the winter solstice, and were eminent patrons of Free-masonry. There is, however, no historical evidence to support this statement, and the celebration of these days by the fraternity generally has been very properly discontinued.
128

## *The Circle Embordered by Two Parallel Lines*

Q. In every lodge may be seen "a certain point within a circle embordered by two parallel lines." Have masons lost the true meaning of this emblem?

A. They have.

Q. What does this emblem signify?

A. The astronomical signification of this emblem is so apparent that it seems to have forced itself upon the attention of many intelligent masons. Dr. Oliver, in his Dictionary, says:

> The symbol of a point within a circle has sometimes been invested with an astronomical reference. Thus it said that the point in the circle represents the Supreme Being, the circle indicates the annual circuit of the sun, and the parallel lines mark out the solstices, within which that circuit is limited. And they deduce from this hypothesis this corollary: that the mason, by subjecting himself to due bounds, in imitation of that glorious luminary, will not wander from the path of duty.

This explanation is concurred in by Dr. Mackey, not withstanding his disapproval of the astronomical theory. It is, however, far more reasonable than the explanation given in the lecture appertaining to this degree, but is not in all respects correct. It is true that the circle represents the ecliptic or annual path of the sun, but the "point within the circle" does not represent the Supreme Being, but the *earth,* around which, as a center, the sun appears to annually revolve among the stars of the zodiac. The parallel lines are the tropics of Cancer (♋) and Capricorn (♑). The summer solstice is on the 21st of June, and the winter solstice on the 21st of December. These are the solstitial points, always marked by two parallel lines representing the tropics, as may be seen on any terrestrial globe or map. These two dates, as we have remarked in the answer to the previous question, have been said to be the respective birthdays of St. John the Baptist and St. John the

Evangelist, but there is no authentic history to substantiate the assertion.

**CIRCLE**
**EMBORDERED**
**BY PARALLEL**
**LINES**

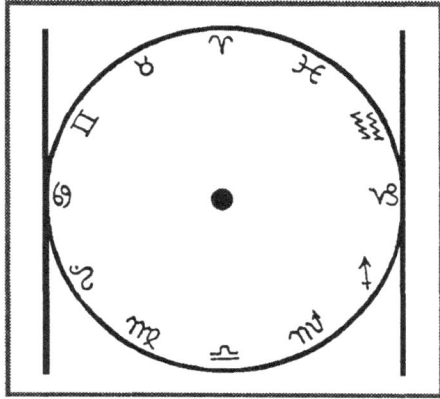

The sun's circuit among the stars is limited and defined by the tropics. When, in June, he reaches the tropic of Cancer, and attains his greatest northern declination he goes no farther north, but turns and begins to approach the south. He reaches his greatest southern declination in December at the other tropic, which terminates his southern progress, for he there again turns about, and once more journeys toward the north. Upon the integrity of the sun's movement, in this particular, depend all the order and regularity of the seasons. Should the sun not retrograde at the summer solstice, the heat would grow intolerable, and both vegetable and animal existence not only become impossible from that cause, but the melting of the polar snows and ice would produce another deluge. If the sun, on the contrary, turned not back at the winter solstice, eternal winter would reign in all lands north of the equator, and a perpetual glacial era extinguish all life and vegetation. The foregoing cut is without question the most ancient and proper method of exhibiting the emblem of "a circle embordered by two parallel lines."

The circle, in order to correspond with *our modern ideas* of the points of the compass, should be turned about so that the two parallel lines would be in a horizontal and not in a perpendicular position. The tropic of Cancer (♋) would then be at the north, and the tropic of Capricorn (♑) at the south, in accordance with our custom of devoting the top of any map or draft to the north and the lower part to the south. The Hebrews, however, and other ancient Oriental nations, when speaking of the four quarters of the heavens, *always supposed the spectator to be looking east,* toward the rising sun; and in their language *"before"* meant "east," *"behind"* "west," the *right* hand south, and the *left* north—as, for instance, the Hebrew word *kedem* means not only *before,* but also *east.* The same custom as to the points of the compass prevailed with the Arabians, who called the north *shemal,* a word meaning *at the left.* This ancient custom fully accounts for the fact that in this emblem the two lines representing the tropics are placed in a perpendicular and not in a horizontal position. The further fact that in some of the Indian cave-temples the circle is found, actually inscribed with the signs of the zodiac, makes the correctness of the foregoing explanation certain. The absurdity of supposing that the operative masons of the middle ages invented this emblem in connection with their art is plain.

Q. Of what does this emblem admonish us?

A. As the sun, in his annual course around the circle of the ecliptic, perpetually performs his revolution with regularity and certainty, never straying beyond the tropical points, but always returning in due season to beautify, adorn, and fructify the earth, so ought we all to govern our actions with equal certainty and regularity, adorning our lives with wisdom and virtue, and making our years fruitful of good deeds, never suffering our passions to lead us beyond the boundary line of good conduct or the points of reason, for, while we keep ourselves thus circumscribed, it is impossible for us materially to err.

## The Lamb-Skin

Q. To what does the lamb-skin, or white-leather apron, allude?

A. The vernal equinox, where the sign *Aries* is found. This sign, as we have seen, teaches immortality, as well as being emblematic of innocence and beauty.

It is a mistake to suppose that the apron, as an article of dress, was confined in ancient times to operative masons and other mechanics. On the contrary, it was an indispensable part of the ordinary apparel of the ancient Egyptians of all classes, and was worn by kings, priests, and nobles, as well as the common people. The apron of the king was, however, of a peculiar form, which belonged exclusively to his rank. It was richly ornamented in front with lions' heads and asps, and other devices, and was of colored leather. The priests, also, wore aprons of peculiar form, as a distinctive part of their sacerdotal dress; so also did the hierogrammat, or sacred scribe. (Wilkinson's "Ancient Egyptians.")

**"LAMB-SKIN"
OR LEATHER
APRON**
(BELZONI)

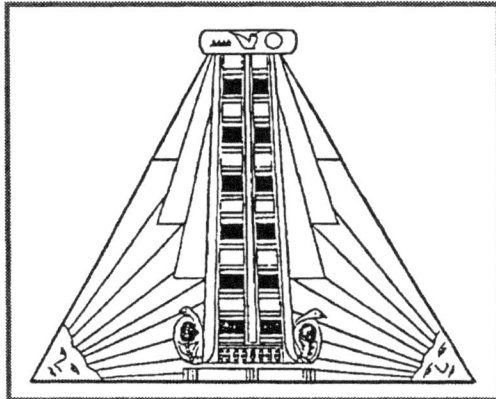

The aprons used in the Mysteries, and by certain sacred officers, were of a *triangular* form, consisting of two parts, as represented above. In the central part the asps are seen, and in the lower corners are lions' heads.

The following drawings are taken from ancient Egyptian monuments. Fig. 1 represents Rameses the Great offering cups of wine in the temple (B.C. 1322). Fig. 2 is the hierogrammat, or sacred scribe. It will be observed that an apron is part of their regalia, each of a different pattern, according to their rank and office.

| FIGURE 1 | FIGURE 2 |

## The All-Seeing Eye

Q. Whence originated the emblem of the all-seeing eye?

A. In most of the ancient languages of Asia, "eye" and "sun" are expressed by the same word, and the ancient Egyptians hieroglyphically represented their principal deity, the sun-god Osiris, by the figure of an open eye, emblematic of the sun, by whose light we are enabled to see, and which itself looks down from the midst of heaven, and beholds all things. In like manner masons have emblematically represented the omniscience of the great Architect of the universe.

**THE EYE OF OSIRIS**
(WILKINSON'S
"ANCIENT EGYPTIANS")

The foregoing is a copy of the Egyptian emblem of the eye of Osiris, taken from the ancient monuments, and found both painted and sculptured on the yet remaining temple walls.

## Masonic Signs

Q. To what does the first sign of an E. A. M. allude?

A. To the autumnal equinox, or place of darkness, and the sign Libra (♎), which is found there, composed of two parallel lines. This sign teaches equality, because at the equinox the days and nights are equal. Equality is the first lesson which a mason receives:

> The king from out his palace must leave his diadem
>     outside the mason's door,
> And the poor man find his true respect upon the check-
>     ered floor.                                            (Morris)

The sign Libra also teaches us to weigh all things in the scales of reason. It is probable that the first sign of an E. A. M. alludes to *both* equinoctial points. When the sun enters Libra he takes the *first* of the *three* leading to his overthrow at the winter solstice; and in like manner, when he enters Aries, at the vernal equinox, he takes the *first* toward his exaltation at the summer solstice.

Q. To what does the first sign of a F. C. M. allude?

A. The first three signs of the zodiac, subtending an angle of 90° from the vernal equinox to the summer solstice.

Q. What does the first sign of a M. M. denote?

A. That all is beneath him, and alludes to the sun, which, when raised into the *third sign* from the vernal equinox to the summit of the zodiacal arch, looks down upon all the signs and constellations beneath him; so, in like manner, a mason having taken the third degree has attained an equal masonic elevation. It also alludes to that benediction or blessing which the sun of the summer solstice bestows upon the labors of the husbandman, and has always been considered the sign of benediction and prayer. (Matt. 19:13-15; Acts 6:6; 13:3.)

## *Masonic Significance of the Zodiacal Signs*

Q. Have the zodiacal signs any further masonic significance?

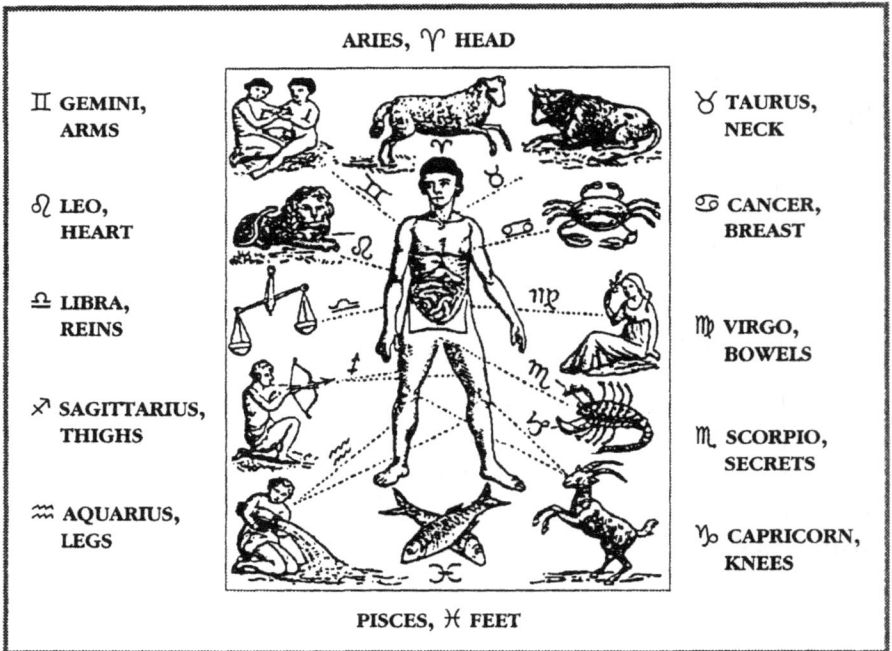

ASTROLOGICAL FIGURE OF *HOMO*

A. They have, of some important particulars: *Astrology* was a leading branch of astronomy as cultivated by the Egyptians. The first six signs of the zodiac, counting from the vernal equinox forward toward the autumnal equinox, may be divided into three parts, typical of the first three degrees of masonry. If we count from the first point of Aries forward 60°, we reach and include Taurus. These two are typical of the first degree, and the unskilled workmen who labored at the preparation of the materials of the temple. Advancing 60° farther, we reach and include Cancer. These two signs, Gemini and Cancer, are emblematic of the second degree. Sixty degrees more take us to and include Virgo, which brings us to the autumnal equinox, typical of completion, and the skilled workmen who wrought at the completion of the temple. The "significators" (to use an astrological term) of the first three degrees may therefore be said to be Taurus, Cancer, and Virgo.

Now, according to the ancient science of astrology, as cultivated by the Egyptians, the sign Taurus (♉) rules the neck and throat; the sign Cancer (♋) the breast; and the sign Virgo (♍) the bowels. The deep and singular significance of this will not be overlooked by any intelligent mason. The astronomical rule of the twelve signs over the various parts of the body, according to astrology, is still kept alive by the figure of *"Homo,"* as seen in old almanacs.

Champollion says the accompanying figure is from the Egyptian Ritual of the Dead, and is often found in their papyri. For further information as to the nature of the rule and influence of the twelve signs, see Lilly's "Astrology." Some very interesting remarks on this subject may also be found in that curious book of Southey's called "The Doctor," Vol. II, Chapter LXXXVII, P. I. See, also, "Sibley's Astrology," Zadkiel's "Grammar of Astrology," and Burton's "Anatomy of Melancholy."

As this division of the first six signs into three equal parts makes *Virgo* one of the "significators" of the third degree, we are naturally reminded of the beautiful virgin alluded to in the modern lecture appertaining to that degree.

### *The Beautiful Virgin of the Third Degree*

Q. What is the origin of the masonic emblem of the beautiful virgin?

A. Although the figure of a virgin is no doubt a very ancient emblem, alluding to the Egyptian goddess Isis and the constellation Virgo as well as to the moon, yet the masonic emblem, as depicted in our monitors, is of late origin. It is in its main features in direct violation with masonic legend and Jewish law and custom. It could not have had an ancient Hebrew origin, for the following reasons:

1. The Jewish law forbids the making of any graven images of the kind. Even the Jews of the present time will not permit any sculptured figures to be set up as monuments in their cemeteries.

2. The urn, which is represented as containing the *ashes* of O. G. M. H. A. B., implies *cremation,* which was contrary to the fixed custom of the ancient Jews, as well as Egyptians, which dictated burial.

3. The Jewish law considered the contact, or near approach even, of a dead body unclean, requiring those thus exposed to undergo a long period of exclusion and purification. Our G. M. H. A. could not, therefore, have been buried anywhere even in the neighborhood of the temple, much less near the *sanctum sanctorum* itself. We have, however, positive testimony as to the modern origin of this emblem. A full history of its invention, and when and by whom introduced into masonry, is given in a late article by Brother Robert B. Folger, in the "Masonic Newspaper." As the communication is of much historical interest, and also fully illustrates the way in which many modern innovations have been made, we give it entire.

It should, however, also be observed that Cross did not claim to have invented all of his hieroglyphics, but admits that many of them had been "described by authors who had gone before him."

FICTION OF THE
WEEPING VIRGIN —
THE MONUMENT OF
HIRAM ABIF
(BY ROBERT B. FOLGER, 33°)

## Fiction of the Weeping Virgin

JEREMY L. CROSS has been dead for many years. A more genial and kind hearted man was not to be found, and his labors in and for the benefit of the masonic fraternity have endeared his memory to all who were acquainted with him during life. He has left a memorial of his masonic labors in the "Hieroglyphic Monitor," which bears his name, which passed through eighteen large editions before his death, and which has been trespassed upon more by masonic work whichever issued from the press, it being the basis of all works of the kind claimed by other persons.

It was my privilege to make the acquaintance of Brother Cross in 1853, at which time he was in the wholesale paper business, in Pearl near John Street, in the city of New York. I became more than commonly intimate with him, and that intimacy increased and continued up to the day of his death. The history of his life, together with all the incidents connected with the publication of his first "Hieroglyphic Monitor," were very frequently the subject of our conversation, and I found that the book was perfectly his "hobbyhorse"; he looked upon

138

it as one of the greatest and most important achievements of his life.

The causes which led him first to devise the plan of such a work were as follows: He was passionately fond of masonry, studied under Thomas Smith Webb, Gleason, and others, became perfect under them in the lectures and work, and then started through the country as a lecturer in the year 1810. He was a man of excellent appearance in early life, strictly temperate from his youth up. His manners were prepossessing, open, frank, very fluent in language, and, withal, a very fine singer. As a matter, of course, he became very popular, the business of lecturing flowed in upon him very fast, and he had as much to engage his mind in that line as he could well attend to. Wishing to take advantage of all the business that offered, he found the work slow of accomplishment by reason of delays caused by imperfect memories. He wanted something of an objective kind, which would have the effect of bringing to mind the various subjects of his lectures, and so fixing the details in the mind as, with the sets of objects presented to the sight, the lectures in detail would be complete.

There was not at that time any guide for lodges except the so-called "Master's Carpet," and the works of Preston and Webb. The "Master's Carpet" was deficient, being without many of the most important emblems, and those which it displayed were very much "mixed up." The work of Preston did not agree with the "adopted work." That of Webb agreed perfectly, but still was wanting in it most important part, viz., the hieroglyphics, by which the work is plainly and uniformly presented to the learner, rendering it easy of acquirement, and imprinting it upon the mind in such a manner that it will not readily be forgotten.

The second object was a copyright. He knew that in those days the cost of bringing together and putting together, and the bringing out of a work of the kind which he desired, would throw him into a large expenditure, and, in order to get back the cost and derive any solid benefit from it in the end, it must of necessity be in his own hands alone.

He considered the matter for many months, and finally attempted to draw various plans, taking Webb's "Monitor" for a guide. Part of the work he accomplished satisfactorily to himself. This included the first and second degrees, and, although there was but little really original in the emblems which he produced, yet the classification and arrangement were his own. He went on with the third degree very well as far as the "Monitor" of Webb goes, when he came to a pause.

There was a deficiency in the third degree which had to be filled in order to effect his purposes, and he became wearied in thinking over the subject. He finally consulted a brother, formerly a Mayor of New Haven, who at the time was one of his most intimate friends, and they, after working together for a week or more, could not hit upon any symbol which would be sufficiently simple and yet answer the purpose. Whereupon the copperplate engraver, also a brother, who was doing his work, was called in. They went at the business with renewed courage, and the number of hieroglyphics which had by this time accumulated was immense. Some were too large, some too small, some too complicated, requiring too much explanation, and many not at all adapted to the subject. Finally, said the copperplate printer:

"Brother Cross, when great men die, they generally have a monument."

"That's right," said Cross; "I never thought of that," and away he went.

He was missing from the company, and was found loitering around the burying ground in New Haven in a maze. He had surveyed all that was there, but did not seem satisfied. At last he got any idea, whereupon the council came together again, and he then told them that he had got the foundation of what he wanted—that while sojourning in New York City he

140

had seen the monument erected over Commodore Lawrence, in the southwest corner of Trinity churchyard; that it was a glorious monument to the memory of a great man who fell in battle. It was a large marble pillar, broken off. The part broken off was taken away, but they had left the capital lying at the base. He would have that pillar for the foundation of his new emblem, but would bring the other part of the pillar in, leaving it to rest against the base. Then one could know what it all meant. The other part of the pillar should be there. This was assented to, but more was wanted. They needed some inscription describing the merits of the dead. They found no place on the column, and after a lengthy discussion they hit upon an open book placed upon the broken pillar. But there should, in the order of things, be some reader of the book; so they selected the emblem of innocence in a beautiful virgin, who should weep over the memory of the deceased while she read of his heroic deeds.

"But, sir," said I, "how will you get along with the Jewish people? You know that very many Jews are masons. They are very tenacious of the 'law' which forbids the making of any image of any kind, and that even the touch of a dead body by a Jew renders him unclean, and, as a consequence, unfit to come into the synagogue until after many days' purification. They would never allow any dead body to be brought into the temple, nor will they even to this day allow any sculptured figures or images to be put up as monuments in their cemeteries."

"Oh, I never thought of that," said Brother Cross. However, it makes no difference. I did not intend to injure the feelings or prejudices of any one by my monument. I only invented it to serve as a help to memorize my lectures and work.

"Admirable, indeed," said I, "but how does it happen that, in the year 1825, when I was raised to the third degree, in Fireman's Lodge, old City Hotel, there was nothing mentioned

[1] Captain Lawrence: see "American Cyclopaedia."

about any monument of the kind! How did it get into the history at all?"

"Oh," said Brother Cross, "I put it there. You see the work was imperfect without the monument. It was right that there should be a monument for great men when dead. The thought of burying the body of a great man without leaving some memorial to mark the place where he is laid is repulsive. I think I have supplied the deficiency, and done it admirably."

"But, still, this was done in 1819, and in 1825 it had not reached New York."

"Oh, that is right. The Grand Lodge of the State of New York would not receive my work, and did not until 1826. They worked 'old style.' All the Eastern, Southern, and Western States had received and authorized it, but New York and Pennsylvania held out. But in 1826 Brother Henry C. Atwood, one of my ablest scholars, and as good a workman as I ever saw, established Mystic Lodge in New York City, and worked after my system. Immediately the work spread throughout the State.

"The craft are indebted to me for harmonizing and beautifying the work and lectures. I have labored solely for their benefit, and they are quite welcome to all that I have done. But many have treated me badly, by copying and publishing my hieroglyphics, claiming them as their own. My copyright was based upon them, and upon the order of their arrangement. The publication cost me a large amount of money, and involved me in debt; and soon after its appearance a lecture in Vermont made a similar publication, infringing upon my copyright. I sought redress from the law, and was sustained. My copyright was confirmed and secured.

"Since that I have never pushed the matter, although frequently on the point of doing so, as all those difficulties generally ended in some compromise, which amounted to very little. Many of the hieroglyphics which I have used are described by the authors who have gone before me, yet there are *many* which are not described, or even made mention of.

These I claim as my own property, and, if I have refused to proceed in law against those brethren who have wronged me, it was not, because I doubted the justice of my claim or my ability to recover. This had been already settled in law. I chose to remember my obligations to the Order, although others had forgotten them. I preferred to dwell in unity and peace with the brethren rather than be the author of contention and strife, and thus bring a reproach upon an institution which I venerate and love."

It would be proper to state that the monument erected to the memory of Commodore Lawrence was put up in the southwest corner of Trinity churchyard, in the year 1813, after the fight between the frigates Chesapeake and Shannon, in which battle Lawrence fell. It was a beautiful marble pillar, broken off, and a part of the capital laid at its base. The monument remained there until 1844-45, at which time Trinity Church had been taken down and rebuilt as it now stands. When finished, all the *debris* was cleaned away, the burial-grounds trimmed and fancifully decorated, and the corporation of the church took away the old and dilapidated monument of Lawrence from that spot and erected a new one of a different form, placing it in the front of the yard on Broadway, at the lower entrance of the church, where it now stands. Brother Cross and myself visited the new monument together, and he expressed great disappointment at the change, saying, "It was not half as good as the one they had taken away."

Brother Cross was a lecturer in masonry for more than forty years, and his name will be cherished by masons for many generations to come. ("Masonic Newspaper," New York, May 10, 1879.)

Below is a view of the Lawrence monument, formerly in Trinity churchyard, referred to in the foregoing article from the "Masonic Newspaper," and from which it said Cross took his emblematic monument of Hiram Abif. (See Lossing, "Pictorial Field-Book of the War of 1812.") It will be observed that the

THE LAWRENCE
MONUMENT

weeping virgin, the open book, and *the figure of Time, are* all
wanting. As these form the essential features of the masonic
monument, Cross must have obtained the most significant elements of his emblem from some other source, which has not
been disclosed.

Had Cross been more familiar with the symbolism of those
ancient Mysteries from which Freemasonry is derived, he
might have devised such an emblem as he desired, which,
while it expressed the same general idea, would not have thus
violated the traditions of our Order, and also, at the same time,
have been in entire harmony with the astronomical basis of
the legend of the third degree.

Among the many names under which the constellation
Virgo was adored was that of Rhea. This goddess was figured
(according to Bryant) as a beautiful female adorned with a
chaplet, in which were seen rays composed of ears of corn
(i.e., wheat), her right hand reclining on a pillar, and in her left
spikes of corn. By corn the ancients intended wheat. *Maize,*
which in America is almost exclusively called corn, was not
known until the discovery of this continent. The spikes of
"wheat" in the chaplet and left hand of the goddess Rhea are,
like those held in the left hand of Virgo, emblematic of the
144

**THE BEAUTIFUL VIRGIN OF THE THIRD DEGREE**

season when the sun enters that sign. This figure of the goddess Rhea, it will be seen, resembles somewhat the virgin of Cross, standing by the broken column, holding in her hand a sprig of *acacia* instead of the spikes of wheat. Rhea was the daughter of Sky and Earth *(Coelus* and *Terra)*. She was also the mother of Jupiter and wife of Saturn, also known as *Kronos,* or Time. This would quite naturally permit the association of the figure of Saturn and his scythe—or Time—with that of the virgin. In the Dionysiac Mysteries, Dionysus (who is the same as Osiris, the personified sun-god) is represented as being slain. Rhea (who is also identical with Isis and Virgo) goes in search of his body, which she at last finds, and causes it to be buried with due honor. Now if, as Dr. Mackey admits, this legend was introduced into the fraternity established by Hiram at the building of King Solomon's temple, and forms the basis of the third degree of Freemasonry, this figure of the goddess Rhea would be a very appropriate emblem of that degree.

Thus the present emblem of the beautiful virgin requires but slight modifications to bring it into entire harmony with all the ancient traditions and mythology. The *pretended history* illustrating the emblem, which Cross admits he invented, should be expunged from the ritual, and the figure of the beautiful virgin represented somewhat after the manner here depicted.

The open book and funeral urn are omitted for the reasons before given. In the *left* hand thus placed at liberty is the evergreen, or sprig of *acacia,* because in her left hand Virgo holds the spear of ripe wheat, for which masons have substituted the former as an emblem of immortality—although to those who are familiar with the beautiful utterances of St. Paul, the spike of wheat is as significant an emblem of eternal life as the evergreen. Says the apostle:

> But some will say, How are the dead raised up, and with what body do they come? Fool, that which thou sowest *is* not quickened except it die, and that which thou sowest is not that body which shall be, but bare grain, it may chance of wheat, or some other.

The *right* hand is represented as resting on the broken column, because the ancients figured Virgo, under the name of Rhea, with her right hand resting on a stone pillar.

The alterations thus made in the emblem are but slight, and nothing is omitted but the "funeral urn" and the "open book." The latter is represented by Cross in a shape entirely unknown to the ancients, whose only books were in the form of rolls of manuscript. The handsome octavo volume, which he has placed on the broken column, looks as if just issued from the press, and is a gross anachronism. Those who are familiar with the lectures belonging to the third degree will find an additional and masonic reason for placing the evergreen in the left hand, "for, as the left is considered the weakest part of the body," it is thus more significant of its mortality: the *acacia,* therefore, placed in the left hand, more clearly teaches us that, when the body, by reason of its weakness, crumbles into dust, the soul of man, rising from the "rubbish" and ruins of its earthly tabernacle, shall dwell in perpetual youth in that "temple not made with hands, eternal in the heavens." Behind the figure of the virgin stands the form of Saturn, or Time, not counting the ringlets of her hair, but pointing upward toward the summit of the zodiacal arch. This beautiful daughter of the skies, Virgo, according to other mythological legends, is also the husband of the sun, who, when he entered the constellation Virgo, was said to espouse her.

The whole emblem may therefore be astronomically explained as follows: The virgin weeping over the broken column denotes her grief at the death of the sun, slain by the wintry signs. Saturn standing behind her and pointing to the summit of the zodiacal arch denotes that *Time will heal their sorrows,* and, when the year has filled its circuit, her lord the sun will arise from the grave of winter, and, triumphing over all the powers of darkness, come again to her embraces.

The emblem of the beautiful virgin, thus represented and explained, is not only an eloquent expression of affection weeping over the loss of a beloved friend, but also a mystic symbol of some of the leading facts of astronomy, and a significant emblem of the immortality of the soul.

### The Evergreen ...

Has been selected by masons as an emblem of immortality, because, when in the icy grasp of winter the whole vegetable kingdom lies dead, it alone blooms in beauty, reminding us of the vernal equinox, when all nature shall revive again:

> ". . . . the evergreen
> That braves the inclement blast,
> And still retains the bloom of spring
> When summer days are past;
> And though the wintry sky should lower,
> And dim the cheerful day,
> It still retains a vital power,
> Unconscious of decay."

## The Sprig of Acacia

Q. Has the sprig of acacia any further signification?

A. The astronomical significance of the *"evergreen, "* which we have substituted for the Egyptian *acacia,* and its allusion to the vernal equinox and the doctrine of immortality, has already been fully explained and illustrated. The symbolism of the acacia is, however, more extended. The acacia grows in Egypt, and is the plant from which gum-arabic is obtained. It is also the *acanthus* of Herodotus and Strabo.

> The thickets of *acanthus,* alluded to by Strabo, still grow above Memphis, at the base of the low Libyan hills. In going from the Nile to Abydos, you ride through the grove of acacia, once *sacred to Apollo,* and see the canal traversing it, as when the geographer visited that city.      (Wilkinson's "Ancient Egyptians," Chapter VI)

The acacia is also a symbol of *innocence.* "The symbolism here," says Dr. Mackey,

> is of a peculiar and unusual character, depending not upon any real analogy of form or use of the symbol to the thing symbolized, but simply on the double or com- pound meaning of the word. For acacia, in the Greek language, signifies both the plant in question and inno- cence or purity of life.    ("Symbolism," Chapter XXVIII)

We think Dr. Mackey is mistaken in this. He does not seem to have been aware, or has overlooked the fact, that one spe- cies of the acacia is a sensitive-plant.

Pliny mentions a *sensitive acacia* about Memphis. One
is now common on the banks of the Nile above Don-
gola (the *Acacia asperata).* The *"Mimosa Lubek"* also
grew of old in Egypt, and the Copt Christians have a
silly legend of its worshipping the Saviour.

(Wilkinson's "Ancient Egyptians")

The peculiar nature of the sensitive-plant has in all ages
excited the wonder and superstition of man, and there is no
doubt that it was the *Acacia asperata,* or *mimosa,* which was
the species of the acacia held as a sacred plant by the
ancients. The word *acacia* is of Greek origin, and to the lively
and poetical imagination of the Greeks this sensitive-plant,
thus shrinking from the touch, was an expressive symbol of
that innocence which in like manner shrinks from the rule
contact of the world—and thus they named it *acacia,* a word
which means *innocence.* It therefore appears that there is a
real and beautiful analogy "between the symbol and the idea
symbolized," and that this symbolism does not "depend simply
on the double or compound meaning of the word" acacia, as
stated by Dr. Mackey; this *sensitive* plant being named "innocence" because it was the natural and appropriate emblem of
innocence and purity.

### The Letter "G"

Q. Is the custom of displaying the letter "G" in masonic lodges
of any great antiquity?

A. That it can not be must appear evident when we reflect
that masonry existed long before the English language. The
letter "G" as displayed in the lodge is, however, a necessary
and appropriate substitute for the *equilateral triangle,* so
prominently used as a sacred symbol by our ancient brethren.

### The Equilateral Triangle

Q. Why so?

A. For two reasons: 1. The triangle is the true significator
of that noble masonic science, *geometry*—since, without a

knowledge of its form and properties, that science is impossible. It was upon the triangle that Pythagoras erected his celebrated and invaluable *"Forty-seventh Proposition."* He is also said to have discovered that the sum of all the angles of any triangle is equal to two right angles. It is more probable, however, that he brought these two propositions, together with a knowledge of the true system of the universe, with him from Egypt, where he went to pursue his studies, and was initiated into the Mysteries.

2. The *equilateral triangle* is also a sacred symbol of the Deity, being the same in its form as the ancient Greek delta, or letter "D." The Phoenician letter "D," as well as the Egyptian, was of a similar form. The equilateral triangle, in the Greek tongue, as well as many other ancient languages, was thus the initial letter of the name of Deity. In the days of Pythagoras we are told that, whenever an oath of unusual importance was to be taken, it was administered on the equilateral triangle, as, by so doing, the name of God was directly invoked. This oath is said never to have been violated. The EQUILATERAL TRIANGLE, therefore, since it is at once the emblem and essence of *geometry,* and the initial letter of the name of Deity, should be seen in the midst of every regular masonic assembly.

## The Compasses

Q. According to an ancient custom, the compasses, as a masonic emblem, whether reposing on the altar or worn as an officer's jewel, should be set at an angle of 60°. What is the reason of this?

A. The reason is principally geometrical. The sacred import of the equilateral triangle has already been explained. Now, as the sum of all the angles of any triangle is equal to two right angles, or 180°, it follows that each of the equal angles of any equilateral triangle is equal to one third of two right angles (180°/3 = 60°), which is 60°.

The compasses being set at 60°, thus allude to the equilateral triangle, and, if the two points were united by a straight line, one would be formed. There can be but little doubt that it was the equilateral triangle itself which our ancient brethren placed upon the altar, since it was upon that emblem their most solemn obligations were taken. In modern times the compasses, set at an angle of 60°, have been substituted. This may have been done purposely, or it may be that, during the dark ages, some of our ignorant mechanical brethren mistook the sacred emblem for one of their working-tools, and that the change was thus brought about. Other mistakes equally as singular, as will be seen in the sequel, were thus made at that period.

The angle of 60° has also an allusion to the zodiac, being equal to two signs thereof, and, if multiplied by the sacred number three, becomes 180°, or the dimensions of the Royal Arch.

Again, if a circle of any size be drawn, a chord of 60° of that circle will be equal to its radius, and the compasses so set will divide the circumference into six equal parts. The points thus made, taken with the one in the center, constitute the mystic number *seven.* The six exterior points, if joined by six straight lines, will form a perfect hexagon within a circle, one of the perfect figures. Or, if we unite these six points in another way, we have the *double equilateral triangle,* in union with the symbol of "a point within a circle."

This was one of the most sacred of all the emblems of Pythagoras, and is also known even to this day through the whole East, and has been there revered for ages, as the SEAL OF KING SOLOMON, by the power of which he bound fast the *genii* and other spirits who rebelled against God. (See "Arabian Nights," and the story of the "Fisherman and the Genius" for an expression of this belief.) If the whole seven points be joined by straight lines, we obtain the figure of a perfect cube within a perfect *sphere*. (See "Historical Landmarks," Lecture V, and notes.) The cube has in all ages been held sacred.

SEAL OF KING SOLOMON

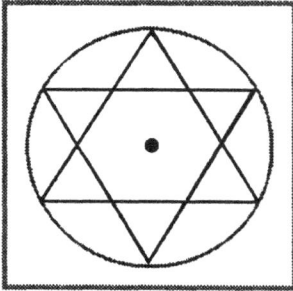

CUBE AND SPHERE

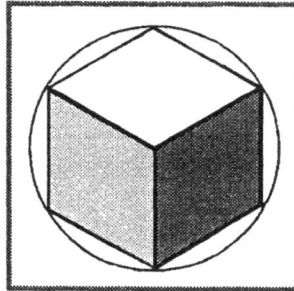

All altars were in the form of a cube, or double cube, which last is the form that ancient custom prescribed for the masonic altar. The ancients esteemed the double cube "holy," but the perfect cube was "most holy." We also read in the Scriptures that the house of God, which King Solomon built, was in the form of a double cube, being *forty* cubits long and *twenty* cubits broad (1 Kings 6). The holy palace itself was a perfect cube, being twenty cubits each way (2 Chron. 3:8). According to the teachings of Pythagoras, also, the cube was the most sacred of all the perfect bodies. From what has been said, the deep emblematic significance of the masonic altar, or double cube, upon which was anciently placed the equilateral triangle, or sacred symbol of Deity, is sufficiently apparent. To this we have in modern times, with great propriety, added, as having a corresponding place upon our altar, the holy Scriptures, the inestimable gift of a later period, the blessing of its possession having been denied to our ancient brethren, from whom, however, was not withheld a knowledge of the true God; but the holy Bible, as we possess it, was not only unknown to Plato and Pythagoras, but also to King Solomon, the wisest of mankind.

## The Emblem of Ears of Corn Hanging by a Water-Ford, or a Sheaf of Wheat by a River

Q. One of the most expressive and beautiful emblems of the fellow-craft degree is the representation of "ears of corn hanging by a waterford," or, as the emblem is also often represented, "a sheaf of wheat suspended near the bank of a river." (See Sickles's Monitor," page 90.) What is the meaning of this emblem?

**SHEAF OF WHEAT HANGING BY A RIVER**

A. Dr. Oliver devotes the whole of Chapter XIX of his "Landmarks" to the consideration of this emblem. It appears that there is, or was, some confusion as to its true meaning. Some old masons seem to think it refers to the *first* passage of the river Jordan by the Israelites under Joshua, when they entered Canaan; at which time the promised land was covered over by fields of ripe corn, which was by them then assumed as a symbol of the PLENTY which gladdened the hearts of the famished Israelites after their forty years' wandering in the desert. Another interpretation of the symbol, which Dr. Oliver gives in full, refers to a passage in the life of Jephthah, recorded in Judges xii, by which we learn that the Ephraimites quarreled with him, A bloody battle followed, and the Ephraimites were defeated. Jephthah took possession of the passages of the

# Stellar Theology and Masonic Astronomy

Jordan to prevent their escape. When any of the fugitives attempted to cross over, they were commanded to say *"shibboleth,"* but, as they could not frame to pronounce it right, and said *"sibboleth,"* they were discovered and slain, to the number of forty-and-two thousand. This latter interpretation Dr. Oliver thinks to be the true one. He says,

> Such is the historical account of the warfare of Jephthah with the Ephraimites, and the reputed origin of the symbol and its interpretation, because the battle took place *in afield of corn near the river Jordan.*

The interpretation which refers it to the passage of the river under Joshua has been generally discarded by masons, and is not countenanced by the masonic lecture as given in America. The other interpretation, which refers this emblem to the battle with the Ephraimites, is, however, also manifestly incorrect, for the following reasons:

1. There is no history of this battle outside of the Bible and Josephus, and neither account makes any mention of the battle having taken place *"in a field of corn."* Josephus does not even mention the use of the word "shibboleth." (See Judges xii, and "Antiquities," Book V, Chapter VI.) The truth is, the statement that the battle "took place in a field of corn" is purely imaginary, and was invented to make out the interpretation, which otherwise would not explain "the ears of corn," which constitute the leading and most expressive feature of the emblem. It is but another instance of an *interpretation* being invented to explain an emblem, the true meaning of which was lost.

2. This interpretation is also clearly incorrect, from the fact that it has no sort of connection with any other part of masonry, or any masonic event or person whatever. It refers to a period long before the building of Solomon's temple, and is utterly out of harmony with the entire system of Freemasonry and all its details.

The fact that the words *"shibboleth"* and *"sibboleth"* occur in the story told in Judges of the cruel and useless slaughter of the defeated and flying Ephraimites, was seized upon, and seems to have induced the attempt to thus explain the lost meaning of this peculiar and striking emblem; but even then it was necessary to invent an addition to the Scriptural narrative in order to account for the "ears of corn," which were otherwise not explained.

### Sibola

Q. What is the probable true meaning of the emblem of "ears of corn hanging by a water-ford," or "a sheaf of wheat suspended near the bank of a river?"

A. A reference to the Eleusinian Mysteries will go far to clear up the matter, and give us the true import of this symbol. The Eleusinian Mysteries were derived from those of *Isis* (see initial chapter), who was known to the Greeks by the name of *Ceres,* and also *Cybele.* Ceres, or Cybele, was the goddess of the harvest, and was represented, like the beautiful virgin of the zodiac, bearing spears of ripe corn. Isis was in like manner, with the Egyptians, emblematic of the harvest season. In the Egyptian zodiac Isis occupied the place of Virgo, and was represented with three ears of corn in her hand.

The Syrian word for an ear of corn is sibola, identical with shibboleth, which the Ephraimites pronounced, more nearly correct, *"sibboleth."* This word also means "a stream of water," and the emblem of ears of corn or a sheaf of wheat near a watercourse, or river, was one of the emblems of the Eleusinian and Tyrian (or Dionysiac) Mysteries. As the word had a double meaning, the picture formed a sort of rebus. The river is the river Nile, the overthrow of which enriched the soil and brought forth the abundant harvests of Egyptian corn, all of which was symbolically represented by the ears of corn hanging by a river. It is also worthy of remark that the name of the goddess *Cybele,* although differing in orthography, is almost identical in sound with *sibola* in some dialects. This mystic word is therefore a triple *pun,* and has a threefold signification:

    a.    An ear of corn;

    b.    A stream of water, referring to the Nile, upon the inundation of which the harvest depended;

    c.    It might be understood as one of the names of the goddess of the harvest.

Hutchinson, a masonic writer of note, admits that the use of the word *sibboleth* was equivalent to an avowal of a profession of the Mysteries, as it implies *ears of corn.* ("Spirit of Masonry.")

How much more perfect and beautiful is this interpretation of the emblem, and how much more in harmony with the moral teachings of our order! The one explanation recalls nothing to the mind but the bloody and brutal butchery of *forty-two thousand* of his fellow beings by Jephthah, the vile wretch who offered up his own innocent daughter as a burnt-offering (see Judges 11:29-40); the other reminds us of the peaceful pursuits of agriculture, and the benevolence of the GREAT CREATOR, who each year brings forth the harvest in due season, and rewards with *"plenty" the* industry of the husbandman. The improbability of the operative masons of the middle ages having invented this astronomical-agricultural emblem is so plain as to require no comment.

# Chapter 8

# *ASTRONOMICAL*
# *EXPLANATIONS*

*(CONTINUED)*

CONTINUING THE EXPLANATION of the relationship between the symbols of Christianity, Freemasonry, and ancient astronomy:

## *The Pillars of the Porch*

Q. In every masonic lodge may be seen two pillars, surmounted by globes. What is the origin of these pillars, and what do they signify?

A. According to the masonic lecture appertaining to the fellow-craft degree, these two pillars represent those which stood before the porch of King Solomon's temple, and are described in 1 Kings 7:15-24; 2 Chron. 3:15-17; Jer. 52:21-22. The description given in the Bible is very minute, and renders it evident that they were made after Egyptian models. The decorations consisted principally, if not entirely, of network, lily-work,

and pomegranates. Speaking of these, the Rev. Dr. William
Smith, in his *Dictionary of the Bible,* says:

> The Phoenician architects of Solomon's temple deco-
> rated the capitals of the columns with "lily-work," that
> is, with the leaves and flowers of the lily, corresponding
> to the lotus-headed capitals of Egyptian architecture.

The same writer also says in the same work:

> The pomegranate was early cultivated in Egypt: hence
> the complaint of the Israelites in the wilderness of Zin
> (Num. 20:5), this 'is no place of figs, or of vines, or of
> pomegranates.' The tree, with its characteristic calyx-
> crowned fruit, is easily recognized on the Egyptian
> sculptures.          (See article "Pomegranate.")

**PILLARS OF
THE PORCH**

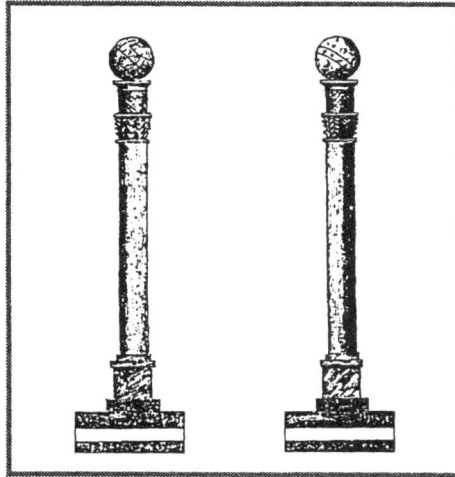

     The    description    of    the
pillars,       as       given       in       the       Bible,       also
renders it probable that they had no globes upon the top of
their capitals, as none are mentioned. This idea of surmount-
ing the pillars with globes arose, no doubt, from a miscon-
struction of the word *"pommels,* "as used in 2 Chron. 4:12-13,
or the word "bowls," in 1 Kings 7:41. That these pommels, or
bowls, were not in any sense academic globes, such as adorn
the masonic columns, is evident from the fact that they were
covered with "network," containing four hundred pomegran-

ates in two "rows, or wreaths" (1 Kings 7:41-42; 2 Chron. 4:12-13). The chapters, of which these pommels or bowls formed a part, were also adorned with lily-work: These pommels must, therefore, have been something entirely different from our modern celestial and terrestrial globes. In place of bearing representations of the "various seas and countries of the earth," and "the face of the heavens," they were *"covered"* by wreaths of network, lilies, and pomegranates. They were not, in fact, globes of any kind, according to Dr. Smith, who says the word pommels "signifies *convex projections belonging to the capitals of pillars."*

The globes that surmount the masonic columns are, on the contrary, modern academic globes, for we find them thus described in the "Monitor":

### *The Globes*

> The globes are two artificial spherical bodies, on the convex surface of which are represented the countries, seas, and various parts of the earth, the face of the heavens, the planetary revolutions, and other important particulars ("Monitor").

It is very evident that no such globes as these could have ever been placed on the top of the pillars of the porch of Solomon's temple even had the sacred text left any doubt upon the subject. Dr. Mackey very truly remarks, in speaking of the symbolical form of the lodge, that "at the Solomonic era, the era of the building of the temple at Jerusalem, the world was supposed to be of an oblong form." Such was the idea held by the most enlightened among the Jewish nation, even down to a very late date, comparatively. Thus, Isaiah (11:12) says, "The Lord shall gather together the dispersed of Judah from the four corners of the earth"; and we find in the Apocalypse (20:9) a prophetic vision of four angles standing on the four corners of the earth. Dr. Mackey, illustrating the ancient idea of the form of the earth (see "Symbolism," Chapter XIII), furnishes a drawing in this form ⬚, within which are marked the "various countries and seas of the earth."

There can be no sort of doubt that such was the prevailing idea of the form of the earth held at that era, not only by the Jews but by most of the other nations. If, therefore, the architect of that age had desired to surmount either of these pillars with a figure representing the earth, he would have placed there a body having the form of a double cube, with the "countries, seas, and various parts of the earth" depicted on its flat upper surface. The same remarks will apply to any representation then made of the "face of the heavens," which, according to the ideas of that age, "was coextensive with the earth taking the same form and inclosing a cubical space, of which the earth was the base, and the heavens, or sky, the upper surface." (Dudley, quoted by Dr. Mackey in note to page 104)

It is, therefore, beyond all question that the introduction of our modern academic celestial and terrestrial globes, as the principal feature and leading ornament of these columns, was not derived from the pillars at the porch of King Solomon's temple. The custom, however, of placing two lofty columns before the porch of temples dedicated to the worship of the heavenly bodies, was a very ancient and universal one. The Egyptian temples were always decorated by such pillars. They may have also ornamented, and probably did sometimes ornament, these pillars with spheres or globes placed on their tops, and intended to represent the one the orb of the sun, or Osiris, the other the full moon of the equinox, or Isis.

That the Phoenician artists who constructed the pillars at the porch of King Solomon's temple also imitated the architecture of the Egyptians in this, is possible, although no mention is made of the fact in either Kings or Chronicles. Such spheres, however, would be something very different from those upon the masonic columns. That the pillars of the porch may have been surmounted by figures globular in form, and intended to represent respectively the sun and moon, is rendered some-what probable from the fact that the whole construction of the temple, as we have seen from what Josephus says, was emblematic of the entire universe. That these columns partook of this symbolism, and were emblematic in some way of the sun and moon, would seem to be indicated by their very names. One of them was called *"Boaz." This* word is derived from two roots, *"bo,"* motion, haste; and *"az,"* fire, i.e., the sun, the great moving fire. The other was called "Jachin," which clearly refers to the moon. Our word "month" is derived from the word "moon"—a month being one moon, or one rev-olution of the moon. The Hebrew months were also lunar, hence they called them *Jachin,* which comes from *Jarac,* which means the moon (Dr. Adam Clarke).

This connection of the globes on the columns, Jachin and Boaz, or the columns themselves, with the moon and the sun, seems to have been at one time fully acknowledged, if not understood, by the fraternity. This connection was no doubt accepted from ancient tradition, while the true cause and real meaning of it was probably lost. The following is a drawing of the two pillars of the porch, taken from a masonic medal struck in 1798, which is but a copy of the way these pillars are represented in the more ancient charts. It will be observed that above the pillar Jachin the figure of the moon is seen, while above that of Boaz the sun appears. (See "Macoy's Cyclopae-dia," article "Medals.")

As to which pillar properly represents Jachin and which Boaz, it must be remembered that, when standing in front of them, they are reversed, Jachin then being on the left hand, and Boaz on the right. In this matter much confusion exists in the pictorial representations made in the Monitors. Kings and Chronicles say that the right pillar was Jachin, and the left Boaz, and the confusion arises as to whether you are supposed to be going into or coming out of the temple. Josephus, how-ever, makes this plain, for, in locating "the table with loaves

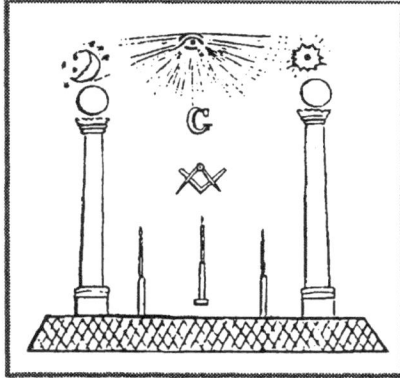

**PILLARS OF THE
PORCH, FROM AN
ANCIENT MEDAL**

upon it," he gives the key to the whole matter, and renders it evident that the pillar Jachin was on the *south* side of the temple, and Boaz on the *north.* He also says the temple itself "fronted to the east." ("Antiquities," Book VIII, Chapter III, and note.) The true position of the pillars is therefore shown by the following diagram:

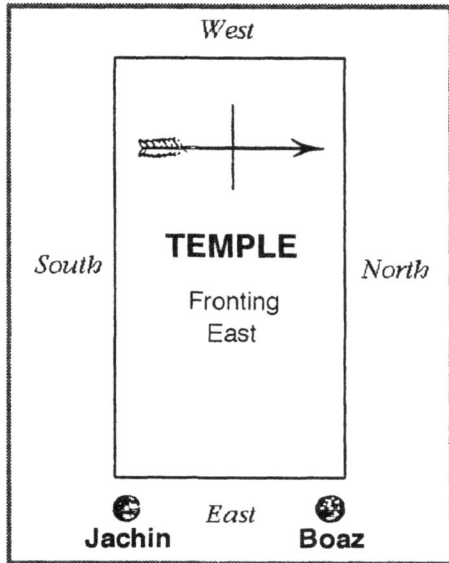

**THE TRUE
POSITION OF
THE PILLARS
OF THE PORCH**

Besides this, the Hebrews, like other ancient Oriental nations, always supposed the spectator looking east, not north, as we do; hence the word *shemal* means *left* as well as *north;* *kedem* means *east,* and also *before;* while the same word which means *south* also means *at the right hand.* When we are told, therefore, in Kings and Chronicles, that the pillar Boaz was on the left side of the temple, it is also implied that it was on the north side. But, as the temple itself fronted to the east, and the pillar Boaz was on the north side of the porch, it also follows that this pillar, which represented the sun, was placed at the *"northeast corner"* of the temple, and in direct line with the rising sun of the summer solstice, as was the case with the ancient temples of Egypt. The full significance of this will be more clearly seen from the answer to the next question, as well as the reason why this pillar was placed on the north side of the porch and not on the south.

It may be thought that, in tracing the primitive meaning of the words Boaz and Jachin to the sun and moon, a conflict arises with what is stated on the margin of both Kings and Chronicles, where Jachin is translated to mean, "He shall establish," and Boaz, "In it is strength." (See 1 Kings 7:21; and 2 Chron. 3:17.) That the words have such a meaning, in a collateral sense, there is no doubt, but the allusion is to the fact that the strength and order of nature, the due course of the seasons, and the division of day and night, were ordained and established by the solar and lunar orbs. "And God said, Let there be lights in the firmament of heaven, to divide the day from the night; and let them be for signs, and for seasons, and for days and years" (Gen. 1:14). The word "strength" is also applied to the sun in many places (see Psalm xix, where the sun is compared to "a strong man, rejoicing to run a race"). The allusion of the words Jachin and Boaz to 2 Sam. 7:16, "And thine house and thy kingdom shall be established for ever before thee" (Simons's "Monitor," page *66),* as given in the fellow-craft lecture, has no foundation other than the fancy of the inventor. The "house" spoken of in Samuel is not the temple, but the royal house, or line of David, just as we now speak of the house of Brunswick, or the house of Hapsburg. It must be remembered, also, that the marginal notes in Kings and Chronicles are really no part of the sacred text, being supplied by the commentators.

The promise made to David is, however, directly alluded to in Psalm 89:35-37:

> Once have I sworn by my holiness, that I will not lie unto David. His seed shall endure for ever, and his throne as the sun before me. It shall be established for ever as the moon, and as a faithful witness in heaven.

Here the connection between the sun and moon, and the ideas of strength and establishment, is directly alluded to, and the symbolism of the pillars of the porch, as representing the sun and the moon, might be appropriately made to refer to the promise made to David. The attempt, however, to make a connection between the marginal notes to Kings and Chronicles, and the text from Samuel, and then to apply them both to the temple, has no foundation in the Bible. The words Jachin and Boaz are simply the names given to these pillars. They mean the moon and the sun, and also strength and establishment, alluding to the respective offices of the sun and the moon. The Hebrew year was lunar, and the moon established years, and months, and weeks; while the sun, "in whom is strength," ruled and divided the seasons. The primitive allusion of the words to the sun and moon is direct. This symbolism, as we have seen by what Josephus says, is in perfect harmony with that which characterized the whole temple, and all parts of it alike. This solar and lunar symbolism of the pillars of the porch was, no doubt, intended to teach the Israelites that the sun and moon were thus to be regarded as emblems only of the great Creator, and not to be worshipped themselves as gods.

As to the globes, if indeed the pillars of the porch were surmounted by globes, the idea must have been derived from Egypt, either directly or through the Tyrian workmen. The allusion of the globes was then, as now, wholly astronomical; but the substitution of our modern academic celestial and terrestrial globes for the orbs of the sun and moon is an innovation of very late date, and was probably the work of Preston, Webb, or, still later, of Cross, author of the "Hieroglyphic Chart," a history of which has been previously given. Cross acknowledges that he invented some emblems, but he also says that many of them had been described before his time. In attempting to depict these, he made many mistakes, from his want of a more intimate knowledge of the symbolism of the ancient Mysteries.

It is true that the Hebrews, and most of the nations at the time of the building of Solomon's temple, did not know the true figure of the earth, yet there is no doubt that the Egyptians were more learned on this point. This, however, while it concedes the Egyptian origin of the globes, does not help the matter, for our *academic* globes, such as are now placed on the pillars, are philosophical instruments of a much more recent date. Apart from this, there can be no doubt that the idea of placing two columns before the temple, however they may have been ornamented was derived from Egypt, where it was the custom, as is not only proved by Herodotus and other historians, but by the temples themselves, remaining to this day, What was the real meaning and true office of these pillars standing before the ancient Egyptian temples, will more fully appear from the answer to the next question.

Brother Robert Macoy, in his "Cyclopaedia," expresses the opinion that the columns Jachin and Boaz were *facsimiles* of the obelisks which stood before the Egyptian temples (see article "Obelisk"). This, of course, does away with the globes, as well as the lily, pomegranate, and network. As to the latter, he is contradicted by Jeremiah, Kings, and Chronicles.

## The Northeast Corner and the Corner-Stone

Q. Why is, or ought to be, the first stone of any building laid in the northeast corner?

A. The ancients believed that the movements, conjunctions, and position of the heavenly bodies influenced not only the destiny of nations, but of individuals, and regulated all the affairs of life. Their temples were dedicated to the worship of the sun, and the whole process of their erection, from the laying of the first stone up to their completion, as well as all the details of the architecture, had special reference to astrological conditions, and the movement of the sun in the zodiac, or his position at stated periods therein.

In our attempt to account for the reason why the corner-stone was laid in the northeast corner, we will, of course, have, in the first place, to resort somewhat to conjecture, as no record of the reason is left; but if by so doing, we finally arrive at a theory, not only in entire harmony with the facts of astronomy, but also with what *is* known of the peculiar customs and religious ideas of the ancients, and which, at the same time, gives a reasonable and sufficient cause, according to the same, for the custom itself, we may feel almost certain that the truth has been discovered.

The cornerstone, we know, was always laid by the ancients with impressive ceremonies and solemn religious rites. As an illustration and confirmation of this statement, the following passage is here transcribed from Tacitus, descriptive of the laying of the cornerstone of the Capitol at Rome, when it was rebuilt by the Emperor Vespasian:

> The care of rebuilding the Capitol he committed to Lucius Vestinus, a man of equestrian rank, but in credit and dignity among the first men of Rome. The *soothsayers,* who were convened by him, advised that the ruins of the former shrine should be removed to the marshes, and a temple raised on the old foundation, for the gods would not permit a change in the ancient form.

On the *eleventh day before the calends of July,* the sky being remarkably serene, the whole space devoted to the sacred structure was encompassed with chaplets of garlands. Such of the soldiers as had names of auspicious import entered within the inclosure with branches from trees emblematic of good fortune. Then the vestal virgins in procession, with a band of boys and girls, whose parents, male and female, were still living, sprinkled the whole place with water drawn from living fountains and rivers. Helvidius Priscus, the praetor, preceeded by Plautius AElianus, the pontiff, after purifying the area by sacrificing a swine, a sheep, and a bull, replacing the entrails upon the turf, invoked Jupiter, Juno, and Minerva, and the tutelar deities of the empire, praying that they would prosper the undertaking, and with divine power carry to perfection a work begun by the piety of man; and then Helvidius laid his hands upon the wreaths that bound the foundation stone and were twined about the cords; at the same time the magistrates, the priests, the senators, the knights, and a number of citizens, with simultaneous efforts, prompted by zeal and exultation, haled the ponderous stone along. Contributions of gold and silver, and pieces of other metals, the first that were taken from the mines, that had never been melted in the furnace, but in their native state, were thrown upon the foundations on all hands. The *soothsayers* enjoined that neither stone nor gold which had been applied to other uses should profane the building. Additional height was given to the edifice, this was the only variation conceded by religion.　　　　　("History" of Tacitus, Book IV, c. 53)

From this it appears that the priests and the soothsayers had the whole control and direction of the ceremony, which was itself of a religious character. This custom was derived by the Romans from a more ancient source, and probably from Egypt, where similar solemn rites were celebrated on like occasions. As all ancient temples were dedicated to the sun primarily, under some of his personal names, we may with good reason believe that the day selected for laying the corner-, or foundation-stone, would be on one of the great solar festivals. Such an occasion would present itself on the arrival of the sun at the tropic at the summer solstice, which indeed would not be far from the *"eleventh day before the calends of July"* mentioned by Tacitus.

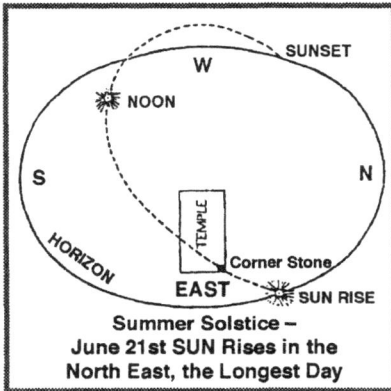

**Summer Solstice –
June 21st SUN Rises in the
North East, the Longest Day**

FIGURE 1

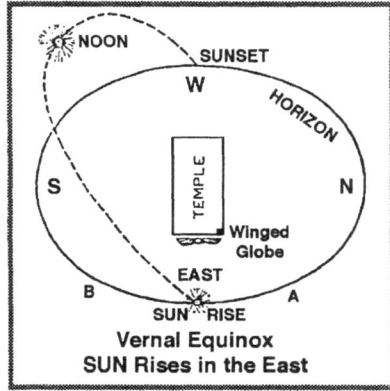

**Vernal Equinox
SUN Rises in the East**

FIGURE 3

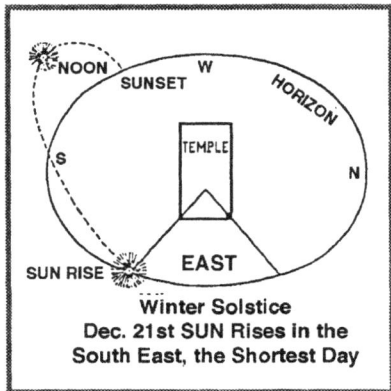

**Winter Solstice
Dec. 21st SUN Rises in the
South East, the Shortest Day**

FIGURE 2

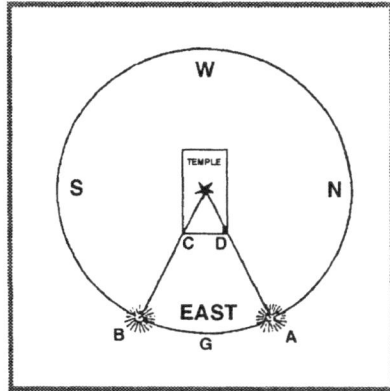

FIGURE 4

**WHY THE FIRST STONE WAS LAID IN THE NORTHEAST CORNER -------
ITS CONNECTION WITH THE SOLSTITIAL AND EQUINOCTIAL POINTS**

The summer solstice was celebrated as a great solar festival by all the ancient nations whose religion had a solar foundation. The day when the sun was thus reached his highest northern declination, and mounted to the summit of "the circle of the heavens," when, according to the teachings of astrology, "he entered his own house" among the stars, would naturally be pronounced propitious and fortunate by the diviners, soothsayers, and astrologers. As the temples always faced the east, so as to catch the first rays of the rising sun, it is almost certain that the cornerstone also, for like religious reasons, would be laid in a line with the rising sun. The sun, as he arose on the longest day of the year, rejoicing in his pride and strength, would thus be a type of the new temple about to rise majestically from its foundations. On the contrary, to lay the cornerstone of the new solar temple in the southeastern line of the sun's decline and fall, at the winter solstice, or toward the north, the point of darkness, or yet toward *Amenti,* the western region of gloom and death, would, according to the teachings of astrology, be most unpropitious, if not sacrilegious.

It therefore of necessity followed that, as the sun on the 21st of June rises in the northeast, and as the future temple itself faced the east, its cornerstone, if placed so as to emblematically represent and mark the place of the rising sun of the summer solstice, must have been laid in the northeast corner. In the preceding diagram Fig. 1 will clearly illustrate this. The dotted line shows the path of the sun from sunrise to sunset on the 21st day of June, or summer solstice. The horizontal circle represents the visible horizon. At this period of the year the sun rises in the northeast and sets in the northwest, as represented by the dotted line, where the respective points of sunrise, noon and sunset are each marked. This drawing also clearly shows the reason why that is the longest day in the year, as it is evident that the circuit from the point of sunrise, by the way of the south to that of the sunset, is greater than at any other time. This custom of laying the cornerstone so as to mark the place of the rising sun of the summer solstice was productive of other useful astronomical purposes; for, due care being taken to establish the proper angle, the *southeast* corner would, as a necessary consequence, be in an exact line with the point of the horizon at which the sun rose at the *winter* solstice. At that period the sun rises in the southeast and sets in the southwest (see Fig. 2). This is the shortest day of the year, for, as will be seen, the path of the sun from the point where it rises to where he sets, by way of the south, is shorter than at any other period. Another consequence followed from this arrangement; for, after the sun quits the south and goes north, when he arrives at the vernal equinox he has journeyed half the distance to the other tropic, and rises at a point due east. At the period of the vernal equinox, the sun rose at a point directly in front of the center of the principal entrance of the temple, which in Egypt *was always surmounted by the sculptured symbol of a "winged globe, " emblematic of the sun,* whose motion was symbolized by the wings.

The same result would also take place when the sun returned from the summer solstice and reached the autumnal equinox. This is illustrated by Fig. 3. The points marked *A* and *B* are those where the sun rises at the summer and winter solstice. It is thus apparent that the porch or front of the temple, from its position and construction, might be used as a perpetual almanac, as the return of the sun to either equinox would be indicated by his rising in a direct line with the "winged globe," sculptured above the principal entrance; and in like manner his arrival at the solstitial points was marked by the northeast and southeast corners of the porch.

The correct marking of the solstitial points in this manner was, however, dependent upon a *certain proportion* (quite easy to determine) *between the breadth of the front of the temple and a point established back of its center,* at such a distance that two lines drawn from that point through corners would cut off the same number of degrees, measured on the horizon, as actually separated the points where the sun rose on the 21st of June and the 21st of December, thus making the front of the temple the *chord* of an *arc* of the same number of degrees

**EGYPTIAN PYLON, OR TEMPLE-GATE, SURMOUNTED BY THE "WINGED GLOBE"**

which separated those two points. The number of degrees contained in this *arc* would depend upon the latitude of the place, increasing in length as we advance toward the north.

In the latitude of Egypt, Rome, Greece and Asia Minor, if this point so established was desired for any reason, to be placed at or near the center of the ground floor of the temple, it would be necessary to build the temple in the form of an *"oblong square"* and in many places the exact form of a *"double cube"* would be required. This may account for the reason why ancient temples were generally built in the form of a "double cube," and why that form was esteemed sacred. This "certain point" back of the center of the front, and in or near the center of the temple proper, might be appropriately marked by an altar, or a *"blazing star"* (emblematic of the sun) *"set in the 'mosaic pavement.'"*

This arrangement, by which the front or porch of an ancient temple was thus made to serve an astronomical purpose, and accurately to point out the commencement of the seasons, is illustrated in Fig. 4. *A* and *B* represent the two points of the horizon where the sun rises at the summer and winter solstice. *C D* represents the front of the temple; the star indicates the point from which the imaginary or actual lines; as the case may be, are required to be drawn; so as to intersect the points *A* and *B,* by passing through the corners of the temple, thus making the front the *chord* of an *arc,* containing the same number of degrees *as A G B*. The other letters indicate the points of the compass. By the use of a "plumbline" a point corresponding to the star might, if required, be established on the roof. This, however, would not be necessary if, as was generally the case, the principal entrances of the temple conducted into an open court, ornamented by rows of pillars. The whole arrangement, if correctly inaugurated by placing the cornerstone in its true position, in the northeast corner, would enable an observer, by use of the most simple and primitive instruments, to determine when the sun reached either of the equinoctial or solstitial points; or, in other words, enabling him to divide the year into its four great natural divisions, and accurately mark the commencement of each.

The length of the solar year could also thus be determined—that is, full as accurately as the ancients did determine it. All of these particulars might, indeed be ascertained without any instruments whatever, by means of *the pillars at the porch.* All ancient temples had two lofty pillars, one at each corner of the porch, and there is no doubt that they had some connection with the arrangement above described. If they wee located with care, the rising sun of the summer solstice would cast the shadow of the northeastern pillar, *Boaz,* along the line *A*★(see Fig. 4), and the rising sun of the winter solstice cast the shadow of the other pillar, *Jachin,* along the line *B*★. Careful observations would also probably be made of the length as well as direction of these shadows at different periods of the year, for at noon on the day of the summer solstice the sun, being higher in the heavens than at any other time in the year, the shadows of the columns would be shortest; and at noon on the winter solstice the shadows would be the longest. These observations of the length of the shadows, being made at noon, would be free from the error occasioned by refraction at sunrise, and thus serve to correct the others.

If these pillars were thus secretly used by the priests for astronomical purposes, it fully accounts for the idea always entertained, but never entirely understood, that the pillars themselves had some connection, actual or emblematic, with the solstial or equinoctial points. The following drawing will clearly illustrate the probable astronomical uses of the pillars of the porch in ancient Egyptian temples. The sun is rising. It is the hour of the morning sacrifice. The pavement of the temple is represented as open to the sky, for the purpose of more easy illustration. It need not have been so in fact, as it is only required that the shadow of the column at sunrise should fall parallel to the solstitial line, which could have been determined from without. In the ancient Egyptian temples, however, the portico and courts leading to the sanctuary were open and uncovered (see Wilkinson's "Ancient Egyptians"), and the shadows of the columns were projected on the floor.

### The Checkered Floor

Whenever it was considered necessary to have the solstitial lines or the length of the shadow actually marked on the floor, then a certain carefully placed line or row of "mosaic squares" (see illustration) would answer the purpose, and also perfectly conceal the design of the whole arrangement; and this is probably the reason why the priests in their temple architecture adopted that kind of pavement. Of course, the details of the arrangement were modified to suit different places and circumstances.

The observations might be made from the roof, or standing in front of the temple, where instruments, simple in construction, for determining the line of direction toward the rising sun, with reference to the front of the temple, might be, and

THE RISING
SUN OF THE
SUMMER
SOLSTICE

probably were, used, without relying wholly on the shadows cast by the columns. The mean daily rate at which the point of sunrise moved along the horizon, and the length of the shadows increased or diminished, would also soon be determined, and thus an observation could be taken at sunrise, noon, and sunset, any day. The month and day of the month could thus be determined at any time with tolerable accuracy. The same arrangement would serve to ascertain the true solar time. Of course, it is now impossible, in the absence of any direct information, to arrive at all the details of the peculiar arrangement by which these ancient solar observations were made, but the main outline is without doubt correct. It was but a more extended application of the principle of the sundial, by means of which not only the hours of the day but the arrival of the sun at the solstitial and equinoctial points, was determined, together with the length of the year and other important particulars. These methods seem clumsy to us, being familiar with the wonderful "instruments of precision" which modern science possesses; but, in those ancient days, such primitive methods were the only ones known, and the accuracy of the results arrived at is a matter of wonder and surprise.

174

The whole arrangement of the porch and pillars of ancient temples for astronomical uses thus depended to a great extent upon the accurate laying of the cornerstone in the northeast corner, so that the outer corner of the same should point to exactly the proper place in the circle of the horizon. The great importance attached to the ceremony of laying the cornerstone is thus accounted for. This explanation, although founded partly on conjecture, harmonizes so well with all that *is* known as to the religious customs and ideas of the ancients, with the facts of astronomical science, and the whole system of solar worship, as to render its truth in the main almost certain. No investigation appears to have ever been made as to the probable connection between astronomy and the porches of ancient temples, beyond the fact, apparent at first sight, that they all face the rising sun; and this is attributed to religious ideas wholly, and not at all to scientific ones, although it was the well-known custom of the ancient priests to conceal the facts of astronomical science under religious allegories.

Those mysterious structures, the pyramids of Egypt, have been more carefully examined, and are found to have been constructed with direct reference to certain astronomical facts, if not uses. The pyramid of Cheops is placed so correctly on the true meridian that the variation of the magnetic needle may be determined by it. It is also so proportional that its height is the radius of a circle whose circumference is equal to the circuit of the Pyramid's base. The long slant tunnel, leading downward from the pyramid's northern face, points to the polestar of Cheop's time.

Professor R. A. Proctor, the astronomer, says in a late article, "The Mystery of the Pyramids" ("Popular Science Monthly Supplement," No. III), that the purpose for which the pyramids were erected "was in some way associated with astronomy, for the pyramids were built with the most accurate reference to celestial aspects." The following is quoted at length from Mr. Proctor's interesting article. We have italicized

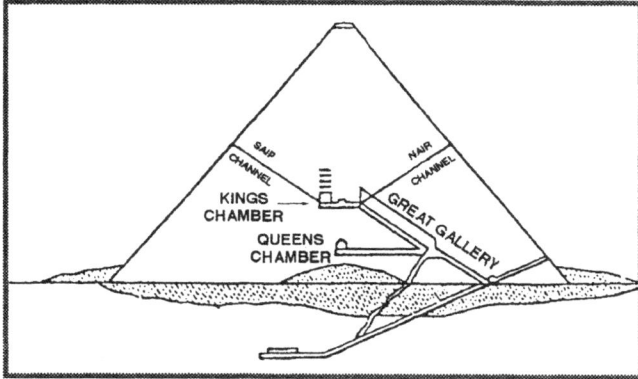

SECTION OF THE GREAT PYRAMID

a line or two as bearing more particularly on our subject.

Mr. Proctor says:

> These buildings [the pyramids] are all, without exception, built on special astronomical principles. Their square bases are so placed as to have two sides lying east and west, and two lying north and south; or, in other words, so that their four faces front the four cardinal points. One cannot imagine why a tomb should have such a position. It is not, indeed, easy to understand why any building at all, except an astronomical observatory, should have such a position. A *temple,* perhaps, *devoted to sun-worship,* and generally to the worship of the heavenly bodies, *might be built that way;* for it is to be noticed that the peculiar figure and position of the pyramids would bring about the following relations: When the sun rose and set south of the east and west points, or (speaking generally) between the autumn and spring equinoxes, the rays of the rising and setting sun illuminated the southern face of the pyramid; whereas during the rest of the year, that is, during the six months between the spring and autumn equinoxes, the rays of the rising and setting sun illuminated the northern face. Again, all the year round the sun's rays passed from the eastern to the western face at solar noon. And, lastly, during seven months and a half of each year, namely, for three months and three quarters before and after midsummer, the noon rays of the sun

fell on all four faces of the pyramid, or, according to a Peruvian expression (so Smyth says), the sun shone on the pyramid "with all his rays."

Mr. Proctor thinks the purpose of the pyramids was rather astrological than astronomical, for he says, "The slant tunnel above mentioned is precisely what the astrologer would require in order to get the *horoscope* correctly." This distinction between astrology and astronomy was unknown to the ancients. The two were one. Astrology assumes, it is true, to predict not only eclipses, but the future generally from the position and aspects of the heavenly bodies; but, in order to make those assumed predictions, it was first required, according to the rules of astrology itself, to obtain a correct knowledge of the position and aspects of the sun, moon, and planets. This necessitated, of course, correct *astronomical* observations, which might be and were put to uses entirely scientific and practical by the ancients, as well as serving as a basis for their pretended predictions of the future.

That the pyramids (whatever else they may have been intended for) were *not* temples, we are perfectly willing to grant, because the only object which has induced this notice of their astronomical proportions, is to show that is a demonstrated fact that the ancient Egyptians did allow the most exact astronomical ideas to greatly influence, if not wholly control, their most stupendous works of architecture—works so gigantic in size, and requiring such an expenditure of time, treasure, labor, and human life, as to render them the greatest wonder of all antiquity. It therefore becomes almost certain that astronomical considerations would not be neglected in the construction of their temples proper, devoted as they were to sun-worship, and the service of a religion having a purely astronomical function.

In ancient times the only astronomers were the priests, and the only observatories the temples. The mass of the people were ignorant and superstitious, and wholly dependent upon the priests for their knowledge required to carry on agriculture. Says Salverti:

> From the observations of the stars, the return of the seasons and several meteorological phenomena were predicted by the priest. He regulated agricultural labors in a rational manner, and foretold its probable success with tolerable exactness. The ignorant men, therefore, under his direction, set no bounds in their own minds to the power of science, and doubted not that the futurity of the moral world, as well as that of the physical, was to be read on the face of the starry heavens. In this mistaken idea they were not undeceived by the priests.

In order to perpetuate these ideas, and so increase and preserve their power and influence, all scientific knowledge was locked up in the sacerdotal order and the Mysteries. Astronomical observations were thus of necessity secretly conducted in the temples, and the methods by which these observations were taken, and the real object of constructions for that purpose, were securely veiled beneath allegorical and religious rites and formulas.

The real and scientific reasons why the cornerstone was placed with such care in the northeast corner having been concealed by the priests, in process of time, when their religion was superseded, were entirely lost. The custom, however was first established under all the sanction and requirements of religion, and came at last to be superstitiously followed, not only as to temples, but all other buildings of any importance, whether built so as to face the east or not. The custom has even descended to this day, which shows that some very important reasons must have led to its adoption in the first place. It is thus that the superstitious observance of this custom required for centuries after the real scientific and the pretended religious reasons for it had not only ceased, but been forgotten.

178

## Chapter 8. Astronomical Explanations (Continued)

### Druidical Temples

That the Druids of Britain celebrated the Mysteries in some form, and had secret symbols and signs known only to those who were initiated into the higher priestly orders, is admitted by all investigators. Nor is there any doubt that the Druidical Mysteries were derived from the Phoenician and Tyrian navigators, who visited that country for tin, and who established colonies there. The principle temple of the Druids was what is now called *Stonehenge,* much of which remains to this day. These ancient remains, it is conceded, were erected by those who worshipped the sun, either actually or symbolically, and the peculiar arrangement of the stones strongly confirms the views we have expressed as to the secret solar significance of the "northeast corner" and "the pillars of the porch." Mr. M. D. Conway, in his "South-Coast Saunterings in England" (and who visited the place), informs us that, some two hundred yards from the entrance of the temple at Stonehenge, there is set up a pillar sixteen feet high. This stone pillar he also says,

> is not only set exactly at that point toward the *northeast* where the *sun rises at the summer solstice exactly over its top,* but has also been set in a place where the ground has been scooped out, so as to bring its top, as *seen from the altar,* precisely against the horizon. Every year people go out on the 21st day of June to see the sun rise above this stone, and that it does so, with *absolute exactness,* admits now of now question.

At the druidical temple at Abury there is a stone pillar in the same astronomical position. These pillars are, it is true, of rough stone, but, had the builders of these Druidical temples possessed the same wonderful skill in architecture as the Tyri ans and Egyptians, from whom their religious ideas are derived, no doubt more elegant if not finely sculptured columns or obelisks would have been erected; nor is it at all strange that the temples built by the rude inhabitants of Britain should be inferior to those of Tyre and Thebes, although erected upon the same astronomical principles for the same symbolical solar worship, since it was far easier to adopt the religious rites and ceremonies of the Phoenicians than to rival their skill in art, or to obtain the services of their architects or artists. It may also be presumed that the Phoenicians themselves, who colonized there in the interests of trade, were more skilled in working the tin-mines, or in commercial pursuits, than in temple building and architecture.

## *The Cornucopia*

Q. Whence was this masonic emblem derived, and what does it signify?

**CORNUCOPIA**

A. The Cornucopia, or Horn of Plenty, is an emblem of purely astronomical origin. It alludes to the constellation *Capricornus*. Capricorn, according to mythology, is the same as *Pan,* or *Bacchus,* who, with some other deities, while feasting near the banks of the Nile, were suddenly set upon by the dredful giant Typhon. In order to escape, they at once all assumed different shapes and plunged into the river—Pan, or Bacchus leading the way. That part of his body which was under water took the form of a fish, and the other part that of a goat. Pan was the god who presided over the flocks and herds. Virgil thus invokes him:

"Pan ovius custos."
"Thou, O Pan! guardian of the sheep."
("Georgics," Book I)

Pan was also the god of plenty. Therefore the *twisted horn* of Capricornus became an emblem of plenty.

According to another astronomico-mythological tale, Jupiter is said to have been suckled by a goat—the meaning of which is that the sun, emerging from the stars of Capricornus at the winter solstice, begins to grow in light and heat as he mounts toward the vernal equinox. He is thus figuratively said to be nourished by this goat. The mythological name of this

PAN

nurse of Jupiter was *Amalthaea.* To reward her kindness Jupiter, it is said, placed her among the constellations, and gave one of her horns to the nymphs who had aided in taking care of him during his infancy. This gift possessed the power of imparting to its holder whatever he desired. On this account

CAPRICORNUS

the Latin word *"cornucopia"* denotes plenty; the word *"Amalthoea,"* when used figuratively, has the same meaning. The whole story is a solar allegory, alluding to the arrival of the sun among the stars of Capricorn, at which time the fruits of the earth—"corn, oil, and wine"—have all been gathered in and stored away, so that, although winter comes to desolate the land, the industrious husbandman is yet blessed with "plenty."

## The Beehive

This was one of the emblems of the Eleusinian Mysteries. The goddess Rhea, according to Bryant, was represented with a beehive beside her, out of the top of which arose corn (wheat) and flowers, denoting the renewal of the seasons and the return of the sun to the vernal equinox.

Q. Whence is the masonic emblem of the hourglass derived?

A. The hourglass was one of the sacred astronomical emblems of the Egyptians. Clement of Alexandria, who gives a description of one of their religious processions, informs us that the *singer went* first, bearing the symbols of music, and that he was followed by the *horoscopus,* bearing in his hand an hourglass, as the measure of time, together with a palm-branch, these being the symbols of astrology or astronomy. It was the duty of the *horoscopus* to be versed in and able to recite the four books of Hermes which treat of that science. One of these books describes the position of the fixed stars; another the conjunctions, eclipses, and illuminations of the sun and moon; and the others their risings and settings. The hourglass is, therefore, peculiarly an astronomical emblem of great antiquity. The *moral* application of this masonic emblem is beautifully given in the "Monitor."

## The Anchor, the Scythe, and the Rainbow

Q. Have the anchor, the scythe, and the rainbow any astro-
    nomical significance?

A. These emblems are only incidentally alluded to in the lec-tures, and have no particular significance as to any part of our ancient rites and ceremonies, except in a general way. They are all of them emblems which have been for ages the com-mon property of all mankind, used either "to point a moral or adorn a tale." The last two are, however, astronomical in their inception, as the scythe appertains to Saturn, and the rainbow is not only a celestial phenomenon, but was also one of the emblems of the Eleusinian Mysteries.

### *The Coffin, Spade, etc.*

These are all common emblems of mortality, and appertain as such to the legends of the third degree. As the astronomical allegory contained in that legend has been fully explained and illustrated, these emblems require no further remark.

## *The Key-Stone, and the Legend of its Loss*

The emblem of the key-stone, as now exhibited, together with the legend of the lost key-stone, would appear to be of very recent date. They belong to the Mark Master's degree, as given to the American rite. The Mark degree, says Dr. Mackey, was taken by Webb from the Scottish rite. Webb, however, "improved the ritual and *changed the legend,* substituting one of his own invention." Another writer informs us that Webb's Mark degree is founded on the European degree of Mark master mason—"the sign, token and sacred sign," of which are exactly the same as the "due guard, real grip, and principal sign" of Webb's degree—although it contains no mention of the *"key-stone, "* but in its stead the *"cubic stone."* The weight of testimony from all sources seems to render it certain that the idea of the "key-stone" and the legend connected with it, as given in the American degree of Mark master, are wholly the invention of Webb.

In making these additions to the legends and symbols of Freemasonry, Webb, however, was under the necessity of making what he added *harmonize* with the principal legend of the third degree, as well as that of the Royal Arch; and, in doing so, he unconsciously rendered his new legend and its accompanying emblem capable of the same astronomical explanation as the original legend, which he desired to thus more fully illustrate. We do not mean to be understood as saying that Webb ever had any such astronomical ideas in his own mind, but, being under the absolute necessity of making the machinery of his new degree harmonize with the really ancient and astronomical legends of the Order, he was unconsciously compelled, by a logical necessity, to render that which he supplemented capable of the same astronomical interpretation as the original and fundamental legends of Freemasonry themselves.

The emblem of the key-stone and the legend of its loss may thus be astronomically explained:

## The Key-Stone

Q. Of what is the key-stone emblematic?

A. Of *strength,* that being the strongest part of an arch, binding the several parts together and thus enabling it to bid defiance to the elements.

KEY-STONE

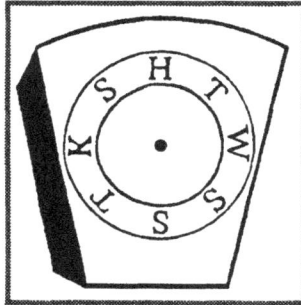

Q. Has the key-stone any astronomical allusion?

A. It alludes to the summer solstice, or key of the zodiacal arch, in close proximity to which it is now seen, and where anciently was located the constellation Leo, also typical of strength.

## The Circle on the Key-Stone

Q. Why is a circle inscribed in the masonic key-stone?

A, A circle is the astronomical sign, and Egyptian hieroglyph of the sun. It is placed in the key-stone to denote the sun in the summer solstice, *exalted* to the summit of the zodiacal arch (see zodiacal figure opposite page 92).

Q. Are the letters surrounding the circle, with the explanation usually given of them correct?

A. As the English language was not spoken at the time of the building of King Solomon's temple, either by the Hebrews or the men of Tyre, the *English* sentence which these letters are said to imply cannot be anything but a very modern innovation. If the circle and its accompanying letters, which Webb placed on his keystone, were borrowed from a more ancient source, he evidently did not know what their true meaning was, and so invented an explanation of his own.

The degree of Master Mark Mason, or Past Master, which was confined to those who had actually presided as masters, while it furnished Webb the groundwork for his new degree, made no mention of the key-stone. It did, however, exhibit the letters H. T. S. T K. S., to which it would appear Webb added a *W.* and another *S.,* for reasons of his own. It is worthy of remark that the meaning attached to these letters has varied considerably. Thus, some fifty or sixty years ago, they were explained as forming the initials of the following sentence: *He That Was Slain Soared To Kindred Spirits,* alluding to the legend of the death of H. A. B., as related in Oliver's Dictionary, and before quoted at length.

The reading of the present day is very different from this, but the reading is not uniform in all the States of the Union. In some States the letters K. S. are said to stand for key-stone, and in others for King Solomon. Some are of the opinion that S. S. stand for *sanctum santorum;* others that the K. T. alludes to Knights Templars. It is evident, however, that there is no definite limit to this mode of reading the mysterious letters; for, proceeding on the same principle, we might suppose them to mean—*Safely Keep This Sacred Secret Within Thy Heart;* or, *Hidden Things We Solemnly Swear To Keep Secret;* or, *There*

*Were Seven Steps To King Solomon's House;* or, *Knights Templars Should Sacredly Watch The Holy Sepulchre;* and thus until our ingenuity or patience is exhausted. And it is also equally evident that all methods of reading these letters, which are founded on the idea that they are in any way initials of English words, must be wrong if the emblem is ancient, and can only be right if it is of quite recent and wholly English or American origin. If, therefore, these letters are of ancient origin, as arranged in this emblem, we may be quite certain that their real meaning has been entirely lost. If they have any ancient meaning, proper investigation and study might, no doubt, rediscover it; but, as we have no evidence whatever that they are ancient, it is not worth while to make any attempt in that direction.

The legend of the loss of the key-stone and its recovery may, however, be brought into harmony with the principal legend of the third degree, and that of the Royal Arch, and thus astronomically explained. All the legends of Freemasonry relating to the finding of that which was lost, refer to the *euresis,* or discovery, by finding of the sun-god, whose death formed the story of the ceremony of the initiation into the Mysteries.

The key-stone is an astronomical emblem of the sun at the summer solstice, or summit of the Royal Arch, after leaving which he is slain, and his body lost among the wintry signs. The astronomical hieroglyph of the sun ☉, which is marked on the key-stone, makes this solar allusion of its loss and recovery perfectly apparent. It may, therefore, be considered as but another allegory of the loss of the sun during the winter months, and his discovery again at the vernal equinox. And, as the name of O. G. M. H. A. B. means the sun, as before explained, the astronomical sign of the sun ☉ on the key-stone is equivalent to his name being there to *mark* or designate the stone as appertaining to him.

### The Legend of the Lost Word

Q. What is the meaning of the masonic legend of the "lost word"?

A. This legend, as briefly stated by Dr. Mackey, in his "Symbolism of Freemasonry" (page 300), is as follows:

> The mystical history of Freemasonry informs us that there once existed a WORD of surpassing value, and claiming a profound veneration; that this word was known to the few, and that it was at length lost, and that a temporary substitute for it was adopted.

This idea of a mystic, all-powerful "word" was an ancient and widely diffused superstition. Just how this notion originated has not been handed down to us, either by tradition or otherwise. It, however, probably came to be entertained in the following manner: It is generally known to the *profane*—i.e., the uninitiated—that those who were admitted to the "Mysteries" were entrusted with a certain sacred word, under a most solemn pledge not to reveal it to the world; and as the scientific knowledge, also secretly imparted to those who were initiated, gave those who took the higher degrees the power to work apparent miracles, the ignorant and superstitious multitude naturally thought, and were perhaps taught to believe, that it was by the use of this "word," so sacredly concealed, that the priests were able to perform all their wonderful works. The word was, however, nothing but the "password" which went with the "sign," by which the initiated could make themselves known to one another. This idea of an all-powerful word was very prevalent among the Jews, no doubt derived from their long stay in Egypt. The notion was that this "word" consisted of the true name of God, together with a knowledge of its proper pronunciation, and that the fortunate possessor of this knowledge became thereby clothed with supernatural power—that by the speaking of this word he could perform all sorts of miracles, and even raise the dead. According to the

Cabalists, "as the very heavens shook, and the angels themselves were filled with terror and astonishment when this tremendous word was pronounced."

Jewish tradition states that God himself taught Moses his true name and its correct pronunciation at the "burning bush." And they believed that Moses, being thus possessed of the "WORD," used it to perform all his miracles, and to confound and overthrow Pharaoh and his hosts. The Jews of a later date, seeking to account for the wonderful works of Christ, asserted blasphemously that he unlawfully entered the "holy of holies," and clandestinely obtained the word used by Moses, which was engraved upon the stone upon which the ark rested. The superstition in relation to a wonder working word also prevailed among the Arabians, who say that King Solomon was in possession of this "grand omnific word," and by its use subdued the *genii* who rebelled against God, many of whom Solomon imprisoned by the use of his magical seal, upon which the word, contained in a pentacle, was engraved. (See the "Story of the Fisherman," and other tales of the "Arabian Nights," where this legend is alluded to.)

It was from these, and other similar legends thus widely diffused among the ancient Oriental nations, that the veneration for a particular word arose, together with an earnest desire to obtain it, and a laborious search for it, by ambitious believers in its power. All the magicians, enchanters, and wonder workers of the East, and the adepts of the West, were supposed to have, in some mysterious way, become possessed of this "word," and were known to the aspirants and students of the occult sciences (not yet so fortunate) by the names of "masters," and the "word" was called by them the "master's word." This ancient superstition seems to have left its impress on our ritual, for the "word," of which we hear so often therein, is assumed to be something more than a mere "pass-word," although we, as masons, now use the phrase

"master's word" in a very different sense from that of the
adepts of former times.

In former and less enlightened times the possession of the
true name of God and its proper pronunciation, or some *substitute for it, authorized by divine command,* were even supposed requisite in order to worship him aright; for it was
ignorantly thought that, if God was not addressed by how
own proper name, he would not attend to the call, nor even
know what the prayers of his worshipers were really
addressed to him, and not to Baal, Osiris, or Jupiter; or, if
knowing, would indignantly reject them. In the East, to address
even an earthly potentate by any other than his own proper,
high, and ceremonious title, was considered both irreverent
and insulting. Among the Jews, however, the pronunciation of
the true name was supposed to be followed by such tremendous effects that a *substitute,* for which they believed they had
the divine sanction, was enjoined. Accordingly, we find in the
Old Testament that, whenever the name of God occurs, the
substitute is used instead of the true name. The word substituted is generally *"Adonai,"* or Lord, unless the name follows
that word, and then *"Elohim"* is used; as, *"Adonai Elohim,"*
meaning, Lord God. From this long continued use of a *substitute* for the real word, the latter, or at least its correct pronunciation, was thought to be lost. A trace of all this is found in
our ritual, and, perhaps, furnishes the true reason why a substitute (as Dr. Mackey informs us in the extract we have
quoted above from his "Symbolism" was adopted.

It will be of no use to trace any further the numerous super-
stitions and legends in relation to this fabled "grand omnific
word." Dr. Mackey very justly says in the work before mentioned, that it is "no matter what this word was, or how it was
lost," for we now know that *no* word can be at present of any
use to a mason, except to serve as a "pass-word," to prove his
right to the honors and benefits of some particular masonic
body or degree; and for that purpose (apart from consider-

ations of a purely *archaeological* and historical nature, one word is just as good as another, so long as it is appropriate to the time and place, and has been established for that purpose, either by ancient usages or some competent authority. Much learning, however, as might be expected, together with persistent search, laborious study, and even the practice of magical arts, have been employed in the past ages, and even down to within a few years, to discover the ancient wonder working word by those who believed in its fabled power, or from a motive of historical curiosity desired to obtain it.

According to some the sacred *Tetragrammaton,* or four lettered name of God in Hebrew, incorrectly pronounced Jehovah, was the true word. Others thought that the Hebrew word *Jah,* the Chaldaic *Bul* or *Bell,* or the Egyptian ON or OM, the Hindu AUM, together with various combinations of them all, constituted the "grand omnific word." But as the possession of no one of them, nor any possible combination of them, seems to confer any miraculous powers on the possessor, neither of them can be the correct one according to ancient traditions. If there ever was such thing as a "grand omnific word" (that is, all-powerful word, from *omnificus* all-creating), it certainly remains lost to this day, and "I fear it is for ever lost," for certainly none of the words disclosed, with so much solemn ceremony, in certain masonic degrees, confer any supernatural powers on those to whom they are communicated.

Q. What astronomical allusion has the ancient legend of the "lost word," as illustrated in the masonic ceremonies?

A. As the masonic legend of the deposit of the "word" in a secure and secret place, and its very consequent loss, has been already quite fully stated by masonic writers, in works sanctioned by the highest masonic authority, there can be no sort of impropriety in relating it here, for the purpose of showing its primitive astronomical significance. The legend is substantially as follows:

Enoch, under the inspiration of the Most High, built a secret temple underground, consisting of *nine* vaults, or arches, situated perpendicularly under each other. A triangular plate of gold, each side of which was a cubit long, and enriched with precious stones, was fixed to a stone of *agate* of the same form. On this plate of gold was engraved the "word," or true-name of God; and this was placed on a cubical stone, and deposited in the *ninth* or *lowest* arch. In consequence of the deluge, all knowledge of this secret temple was lost, together with the sacred and ineffable or unutterable name, or ages. The lost word was subsequently found in this long-forgotten subterranean temple by David, when digging the foundations for the temple, afterward built by Solomon his son.

Other versions of this legend ascribe the building of the underground temple, and the deposit therein of the "word," to Solomon, and its discovery to those "who dug the foundations of the second temple on the same spot, and connect it with the 'substitute ark' deposited in the same place."

Both legends, however, agree in stating that the "word" was buried deep underground, and in the *ninth* arch, or lowest of them all; that it was lost, and remained "buried in darkness" until it was subsequently found and brought to light.

In ancient times, and according to the mystical theology of those days; God and the sacred name of God were supposed to be one and the same. The "word" was itself considered to be, in some sense, a living, creative power. Thus Plato taught that the divine *"logos,"* or word, was God. But, as we have shown, the sun was by the ancients universally adopted as the symbol of God, and subsequently confounded with God, so that the various names of God became also solar names. The loss of the solar name, therefore, became but another expression of the loss of the sun, or sun-god, in the lower hemisphere. Now, let us see how this will harmonize with the legend just related. The sun, having reached the summit of the zodiacal arch, at the summer solstice, begins to descend

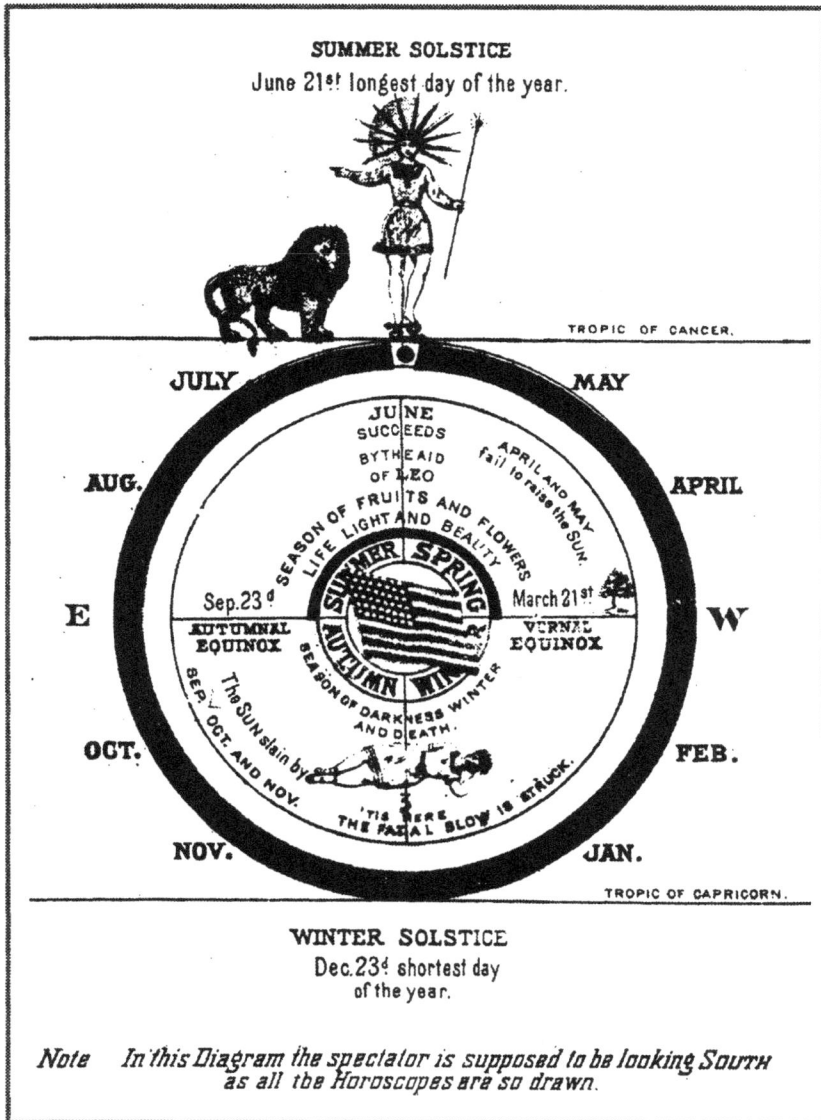

SUMMER SOLSTICE
June 21st longest day of the year.

TROPIC OF CANCER.

JULY      MAY

AUG.      APRIL

JUNE SUCCEEDS BY THE AID OF LEO

APRIL AND MAY fail to raise the SUN.

SEASON OF FRUITS AND FLOWERS LIFE LIGHT AND BEAUTY

SUMMER   SPRING

Sep. 23ᵈ

AUTUMNAL EQUINOX

March 21st

VERNAL EQUINOX

AUTUMN   WINTER

SEASON OF DARKNESS WINTER AND DEATH

E      W

The SUN slain by SEP., OCT. AND NOV.

'TIS HERE THE FATAL BLOW IS STRUCK.

OCT.      FEB.

NOV.      JAN.

TROPIC OF CAPRICORN.

WINTER SOLSTICE
Dec. 23ᵈ shortest day
of the year.

*Note   In this Diagram the spectator is supposed to be looking SOUTH as all the Horoscopes are so drawn.*

toward the region of darkness. From *Cancer* he descends to *Leo,* from *Leo* to *Virgo,* from *Virgo* to *Libra,* and so on until *Capricorn* is reached, which is the *ninth* sign from the vernal equinox, and the undermost of one of the zodiac, corresponding to the *ninth* or lowest arch of the secret vault, and there on the 21st of December, at his lowest declination, at the winter solstice, he is lost, and "lies buried in darkness," until, reviving, he commences his ascent toward the vernal equinox, and begins by his more potent rays to *rebuild* that glorious temple of light and beauty, adorned by flowers and fruits, which the rude assaults of winter have destroyed.

Another allegorical correspondence is found in the fact that the discovery of the word is made according to the masonic legend, by "three," which agrees perfectly with the number of sings, *Aquarius, Pisces, and Aries,* and the months January, February, and March, which separate the winter solstice from the vernal equinox, when, according to the legend of Hiram, the sun is found, as before explained.

The sacred name was engraved on a triangular plate of *gold,* which, according to astrology, is the *solar* metal. This triangular plate, according to the Royal Arch legend, was surrounded by a circle. This triangle within a circle would therefore correctly represent the diagram of the Egyptian *year,* as shown on page 21. But, again, the legend informs us that this triangular plate of gold was fixed to a stone of *agate* of the same form. Now each month, the ancient astrologers taught, had its appropriate gem:

TABLE 2.

| | |
|---|---|
| *Jan.,* the Garnet. | *July,* the Ruby. |
| *Feb.,* the Amethyst. | *Aug.,* the Sardonyx. |
| *March,* the Bloodstone. | *Sept.,* the Sapphire. |
| *April,* the Diamond | *Oct.,* the Opal. |
| *May,* the Emerald | *Nov.,* the Topaz. |
| *June,* the Agate | *Dec,* the Turquoise. |

The *agate,* therefore, is emblematic of the month of June, the summer solstice, and the resurrection and exaltation of the sun. The whole was placed on a *cubical* stone, but the cube was sacred to Apollo, who is identical with Helios, the sun-god. The altar of Apollo at Delos was in the form of a cube. The symbolism of this legend is therefore perfect in all of its details—the emblematic correspondence is too absolute to be accidental. The legend of the lost word is but another form of the solar allegory of the death and resurrection of Hiram, and teaches the same lesson.

### The Masonic Ark

The ark was one of the principal features of the Egyptian Mysteries. Speaking of the religious ceremonies of the ancient Egyptians, Wilkinson says:

> One of the most important ceremonies was the "proces-sion of shrines," which is mentioned in the Rosetta Stone, and is frequently represented in the walls of the temples. The shrines were of two kinds, the one a sort of canopy, the other an ark, or sacred boat which may be termed the great shrine. This was carried with great pomp by the priests, a certain number being selected for that duty, who supported it on their shoulders by means of long staves passing through metal rings at the side of the sledge on which it stood, brought it into the temple, where it was placed on a stand or table, in order that the prescriber ceremonies might be per-formed before it. The same *is* said to have been the cus-tom of the Jews in some of their religious processions as in carrying the ark "to its place, in the oracle of the house, to the most holy place," when the temple was built by Solomon.
> (1 Kings 8. See "Ancient Egyptians," vol. i, page 267)

Wilkinson also says in his notes to "Herodotus,"

> The same mode of carrying the ark was adopted by the Jews (Joshua 3:12; 1 Chron. 15:2, 15; 2 Sam. 15:24; 1 Esdras 1:4), and the gods of Babylon as well as of Egypt were borne and "set in their place" in a similar manner (Is. 46:7; Baruch 4:4-26). Some of the sacred boats, or arks, contained the emblems of life and stabil-ity. which, when the veil was drawn aside, were partly

seen, and others contained the figure of the divine spirit *Nef,* or *Nou,* and some presented the sacred beetle of the sun, overshadowed by the wings of the two figures of the goddess of *Themi,* or Truth, which calls to mind the cherubim of the Jews."

("Ancient Egyptians," vol. 1, page 270; also, note to Rawlinson's "Herodotus," Book II, Chapters LVIII, LIX)

The following drawing is taken from Wilkinson's book and represents the Egyptian ark, with the "sacred beetle" overshadowed by the wings of the double goddess of Truth, copied from the walls of an ancient Egyptian temple.

**EGYPTIAN ARK**
(WILKINSON)

The principal difference between the Jewish and Egyptian arks is that the Egyptian was more like a "boat" in shape, according to our ideas of a boat, while the Jewish ark is described as being of an oblong-square form; this, however it may be observed, was the exact form of Noah's "ark," as described by the Jewish Historian in Gen. 6:14-16. The idea of a boat is therefore characteristic of both of these ancient emblems, as, indeed the very name "ark" denotes.

The above is another view of the Egyptian ark of Osiris, taken from Kitto's "Cyclopaedia of Biblical Literature." The heiroglyphics on the side of the ark are the emblems of dominion, stability, and life everlasting, arranged by 3 x 3.

This mysterious ark, or chest, which figured in the Mysteries of Egypt, much more nearly resembled the Jewish ark in form. After Typhon had slain Osiris,

**ARK OF OSIRIS**
(KITTO)

he inclosed him in a *chest* and cast him into the sea, thus plunging all heaven in grief and sadness. Isis, when she learned the melancholy news, refused all consolation, despoiled herself of her ornaments, cut off her tresses, robed herself in the habiliments of mourning, and wandered forth through the world. Disconsolate and sorrowful, she travelled into all countries seeking the mysterious chest which contained the body of the lost Osiris. In the meanwhile, the chest was drawn ashore at Byblos, and thrown into the center of *a bush,* which, having grown up into a beautiful tree, had entirely inclosed it. At length, however, the tree was cut down by a king of that country, and used by him in the construction of a new palace. But Isis finally learned the singular fate of the chest, and her persevering love was rewarded by the possession of it.

("Philosophical History of Secret Societies," by Rev. Augustus C. Arnold)

The plant which thus indirectly led to the discovery of the mutilated body of Osiris was held sacred by the Egyptians. The whole story of the death of Osiris and the finding of his body is admitted to be an astronomical allegory of the death of the sun-god, slain by Typhon when the sun was in Scorpio, which was at that time on the autumnal equinox. Plutarch informs us that when the sun was in Scorpio, in the month of *Athyr,* the Egyptians inclosed the body of their god Osiris in an *ark,* or *chest,* and during this ceremony a great annual festival was celebrated. Three days after the priests had inclosed Osiris in the ark, they pretended to have found him again. The death of Osiris was lamented by them when the sun, in Scorpio, descended to the lower hemisphere; and, when he arose at the vernal equinox, then Osiris was said to be born anew.

The use made of the ark, or sacred chest, in certain masonic degrees, derives no one of its particulars from anything narrated in the Bible; on the contrary, it bears so striking an analogy to the ark of the Egyptian Mysteries as to at once disclose the original from which it was copied. The masonic ark, like that of the Egyptian Mysteries, is lost or hidden, and after a difficult search at last found. The masonic ark, it is true, does not, like the Egyptian one, contain the body of the slain sun-god Osiris. It does however, contain something symbolically representing the true God, and also certain matters which, it is claimed, lead to a superior knowledge of him. The analogy is therefore perfect, and the astronomical allegory is strictly preserved.

Q. What is the meaning of the emblem of the key?

**KEY,**
**EMBLEM OF**

A. This is a very ancient emblem, and formerly alluded to in the initiation into the Mysteries, which at once unlocked to the aspirant all the hidden secrets of religion, and furnished him with a key to those allegories and tales under which the sublime facts of astronomy and other sciences were concealed from the profane. In Freemasonry it is, more properly, an emblem of the first degree, which, in like manner, furnishes the candidate with a key, and opens the door to the "hidden mysteries of Freemasonry." It has, however, been diverted to the Royal Arch degree, and invested with a different meaning.

**ANCIENT EGYPTIAN
IRON KEY**
(FROM THEBES)

The preceding is a picture of an ancient Egyptian key, from Thebes, and will give a correct idea of the ancient emblem appertaining to the mysteries.

## The Lion, the Eagle, the Ox, and the Man

Q. That is the astronomical allusion of these four ancient
emblems, and why are they thus associated together?

A. They refer to the four great angles of the heavens, where the equinoctial and solstitial points are situated, and the signs at these points are, according to ancient astrology, called "fixed signs." Each sign, was, moreover, ruled by three gods, called *Decans,* the first of which in each sign was called "the power-ful leader of three." The most important and powerful of these thirty-six celestial gods were the four *Decans,* who ruled the four angles of the heavens, and the stability and perpetuity of the universe were supposed to be insured by them. They were also called *Elobim,* and the two who had their seat on the equator were believed to compel the sun to shine twelve hours over all the earth, as well as to repel him, so that he moved on to the next sign of the zodiac in progressive order. The no less powerful Elohim, or Decans, who ruled the solstitial points caused the sun to turn back at the tropics, and preserved the order of nature and of the seasons.

In all ancient astrological projections of the heavens, the four great angles of the zodiac, where these celestial gods were seated, were marked by the figures of the lion, the eagle, the ox, and the man—the constellation *Leo* being anciently at

198

the summer solstice; *Aquarius,* depicted as a man pouring water from a jar, at the winter solstice; and *Taurus,* the Ox, or Bull, at the vernal equinox; while the other angle, or autumnal equinox, was marked by a flying eagle. The quadrants of the celestial sphere were also anciently occupied by the four bright stars *Aldebaran, Regulus, Antares,* and *Fomalhaut.* These were called "royal stars," and in them the four great Elohim were believed to dwell. To them divine honors were paid and sacred images erected, in which the lion, the eagle, the ox, and the man were variously combined. These emblems were worshipped by all ancient nations. The priests and the initiated knew them to be nothing more than astronomical allegories, emblematic representations of the zodiac, but the superstitious people adored them as real gods. The Jews obtained these four emblems from Egypt. Moses, however, forbade their worship, and taught the Israelites to use them to denote the points of the compass and the divisions of their camp, by means of banners on which they were pictured (Numbers 2). These celebrated emblems are therefore of a purely astronomical and zodiacal origin, and, when properly understood (as they were by the initiated), teach many of the most important facts of astronomical science.

### The Royal Arch Banner

Q. What is the meaning and origin of the device on the Royal Arch banner which is represented below?

A. The center of the device consists of the figures of the lion, eagle, ox, and man, the meaning of which has just been explained. The cross which divides them is a correct representation of the equator, cut at right angles by the great solstitial colure, The grotesque and imaginary creatures standing on each side are also astronomical emblems, being compounded of the three figures of the man, the eagle, and the ox—exhibiting the face and body of a man, the wings of the eagle, and the feet of the ox—emblematic of the winter solstice and the autumnal and vernal equinoxes, as before explained. Some are, however, of the opinion that the lower parts of the figures represent the legs of a goat instead of the ox. This would make them refer to Capricornus, the Goat, which *now* marks the winter solstice, thus clearly denoting the "precession of the equinoxes," in consequence of which the figure of the man *(Aquarius)* was changed into that of a goat *(Capricornus),* as the solstitial point left *Aquarius* and entered *Capricornus.*

**ROYAL ARCH BANNER**

Capricornus is also identical in mythology with Pan, who is represented as a god, with the body of a man and the legs of a goat. Astronomical emblems and figures similar to these compound creatures on the Royal Arch banner were quite common among the sun-worshipping nations of antiquity, and were called *sphinxes.* The Egyptians, who held the constellation *Leo* in especial reverence, more frequently combined the human figure with that of a lion, to which they sometimes added the wings of the eagle. These were called *andro-sphinxes;* others, called *crio-sphinxes,* had the head of a ram, alluding to the sign *Aries.* The winged Greek sphinxes, common vases, were partly Egyptian and Phoenician. The Assyrians more particularly esteemed the constellation *Taurus,* and therefore generally combined the figure of a bull with the head and face of a man, to which the wings of the eagle were always attached.

In the Assyrian Museum at the Louvre, M. Botta deposited a slab taken from the palace of Khorsabad, which is ornamented with figures almost identical with those on the Royal

Arch banner. They have a human head, the wings of an eagle, and the legs and feet of an ox. The heads of these Assyrian sphinxes only differ from those of the banner, in being covered by the characteristic Assyrian headdress, and wearing the long ornamented Assyrian beard. Layard also found among the ruins of Nimroud, sculptures of monsters with the head of a lion, the body of a man, and the feet of a bird, which is but a different combination of the same figures, expressing the same astronomical ideas. It is, therefore, evident that the Royal Arch banner is composed wholly of ancient astronomical emblems. The motto, *"Holiness to the Lord,"* is but a proper expression of adoration to the great Creator of the starry heavens, which are so graphically represented by the whole device.

## *The Number "Seven"*

Q. Why was the *number seven* held in especial reverence by
   all the nations of antiquity?

A. The mystic number seven was held sacred by our ancient brethren for reasons which had a purely astronomical origin. The reasons for this will lead us to inquire into the origin of the division of time into days, weeks, months, and years. We were naturally induced to divide our time into periods called days, because the sun makes his apparent diurnal revolution in that time. The Egyptians used to watch for the heliacal rising of the dog-star (Sirius), which, like a faithful guardian, gave notice of the approaching inundation of the Nile, a period of the greatest importance to them, as their harvests depended upon it. By this means a definite period of time was marked off, corresponding to the apparent revolution of the sun in the zodiac. This period was denominated in a *year,* a word which, in our language and all northern tongues, whether *"gear, " "jaar, " "jaer, "* or as in the Persian, *"yare, "*signifies a circle. In Latin, also, the words *annus,* a year, and *annulus,* a circle, are synonymous. Thus the very word "year" alludes directly to the great circle of the zodiac, and points out the origin of that division of time. This period was further divided by the revolutions of the moon about the earth. These latter subdivisions were naturally called "moons," from which is derived our word *"month.* "Among the ancient Egyptians the hieroglyphic sign for a month is the crescent of the moon. In the Hebrew the same intimate connection between the words *moon* and *month* exists as in English. It was also still further observed, by these early students of the skies, that in each lunar month that planet assumed in regular order, at fixed periods of *seven days* each, four distinct phases—the new moon, the first quarter, the full moon, and the last quarter, the "month" was therefore divided into four equal parts of *seven* days each, called *weeks.*

All our divisions of time, whether of days, weeks, months, or years, have therefore an astronomical origin, and are but measures of the observed motions of the moon, for the year itself was originally lunar, the *solar* year having been subsequently adopted on account of its greater accuracy and convenience. The moon, among the nations of antiquity, was the object of universal adoration. Next to the sun in beauty and splendor the moon leads all the hosts of heaven. It may be that the awful majesty and solemn silence of that starry vault, in the midst of which she is seen, caused her to appeal more strongly to the imagination of the early Oriental nations than even the meridian sun itself. It is certain, however, that from ancient Egypt to the distant plains of India, or those far-off lands where the Incas ruled, altars were erected to the worship of the moon, and the goddess adored under a multitude of names, with rites as splendid and awful as those instituted in honor of the sun.

As on every seventh day the moon assumed a new phase, therefore on every seventh day a festival to *Luna* was celebrated. The number *seven* was thus sacred because it was dedicated to the moon. The day set apart for the worship of the moon was known among most northern nations as *"moon-*

*day"*—whence is derived our name for the second day of the week, Monday. The first day of the week being in like manner set apart to the worship of the sun, called *"sun-day."* In fact, each day of the week was set apart to the special worship of some one of the heavenly bodies: Sunday to the sun; Monday to the moon; Tuesday to Mars; Wednesday to Mercury; Thursday to Jupiter; Friday to Venus; and Saturday to Saturn. A strange reminiscence of this fact is found in the modern names of all the names of the week, each of which, like Sunday and Monday, has derived its name from the planet or god to which it was anciently sacred.

Tuesday is derived from the Scandinavian name of *Mars.* The name of the day in French is *Mardi,* derived directly from the Latin, and meaning *"Mars's day"*

Wednesday is from the Scandinavian Mercury, *Woden;* hence *Woden's* day, or Wednesday. The French name of this day is *Mercredi,* from the Latin, meaning "Mercury's day."

Our Thursday is from the Scandinavian Jupiter, *Thor;* hence "Thor's day," and Thursday. The German name is *Donnerstag,* meaning the "Thunderer's day," in allusion *to Jupiter Tonans*. The French call it *Jeudi,* meaning "Jupiter's day."

Friday is named after the Scandinavian Venus, *Fria.* The German name is *Freitag,* with the same derivation and meaning. The French call this day *Vendredi,* which means "Venus's day."

Saturday is derived from Latin, and means "Saturn's day."

The days of the week may, therefore, be just as well designated by the planetary signs as by their names; thus—

TABLE 3.

| ☉ Sunday | ♃ Thursday |
|---|---|
| ☽ Monday | ♀ Friday |
| ♂ Tuesday | ♄ Saturday |
| ☿ Wednesday | |

It was thus that not only the mysterious changes of the moon and the number of the planets, but also the number and order of their religious festivals, and the whole system of ancient worship, were inseparably and astronomically connected with the number *seven* and "the moon, whose phases marked and appointed their holy days." (See Cicero, in the "Tusculan Disputations," Book I, Chapter XXVIII.) It is, therefore, a matter of no wonder that this number should have been held in especial reverence by all the nations of antiquity, or that their imagination should have clothed it with mysterious and magical virtues. This veneration for the number seven was diffused as widely as the worship of the heavenly bodies. The moon was adored in all lands alike, and all her motions, especially her weekly phases, observed with superstitious reverence. It thus happened that, from similar reasons, the number seven was alike considered sacred by nations who had no intercourse, the idea being a spontaneous growth from common astronomical causes.

## The Word "Seven"

The meaning of the *word* seven is also indicative not only of the lunar origin of the division of time into periods determined by the phases of the moon, but also of the universality and identity of the ideas attached to the number itself. The Hebrew word *schiba,* seven, signifies *fullness,* or *completion.* In the Saxon, Persian, Syrian, Arabic, Phoenician, and Chaldean, the word seven has the same signification, and without doubt refers to the moon, which *"fills,"* or becomes *"complete,* "seven days. It is easy to see how a word signifying "filled," or "completed," should be adopted to mark the time when the moon should reach its "full." Before that time she had been increasing in size and light, but now she is filled, or completed; and so, by analogy, the same word in time was also used to mark each period when the other equally distinct, phases of the moon reached maturity.

### The "Figure" Seven

Our figures, 1, 2, 3, 4, etc., are called the Arabian numerals because we derive them from the Arabians, who, it is thought, received them from India. Their true origin is lost in the dim light of extreme antiquity. It is, however, probable that, like the zodiacal and planetary signs, they were originally hieroglyphs. Now, as each seventh day, when the moon assumes a new phase, she has traversed just one quarter of her orbit, we might naturally expect that the hieroglyphic representing the word "seven" would, in harmony with the ancient method of writing, be "a *right angle,* 90°, or one fourth part of a circle." And so, indeed, we find it to be, with only such slight variation as would necessarily result from a constant use of ages, after its emblematic meaning was lost, and only its arbitrary signification was retained. For illustration, let this ⌐ be the original hieroglyph, denoting a period of a quarter revolution of the moon, 90°, and indicating that the moon has "filled," or "completed," *schiba* (seven), one of her phases. The change from ⌐ to 7 is but slight; is but the natural result of the difficulty of rapidly, and without instruments, making a *correct* right angle by the union of two perfectly straight lines, while the lines becoming slightly curved only tended to give the character a more finished and graceful appearance.

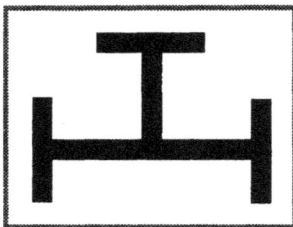

**TRIPLE TAU**

*Triple Tau*

This emblem is not adopted in American Freemasonry, but, placed in the center of a triangle and circle, both emblems of the Deity, it constitutes the jewel of the Royal Arch as practiced in England, where it is so highly esteemed as to be calledthe "emblem of all emblems," and the "grand emblem of Royal Arch Masonry."

The original signification of this emblem has been variously explained. Some suppose it to include the initials of the Temple of Jerusalem, T. H. *(Templum Hierosolymoe);* but, as the *tau cross* as an emblem is much older than the Temple of Jerusalem, this can not be correct; besides, no other evidence is offered for this solution than that the letters T. H. stand for the words *"Templum Hierosolymoe."* We might just as well conclude that the letters stand for "Thrice Holy," "Hiram Tyrian," or the name of any other thing for which the letters T. H. or H. T. may be the initials. Neither is any proof offered to show that the emblem is really composed of the letters T and H, instead of three tau crosses united. Others say it is a symbol of the mystical union of the Father and Son, H signifying Jehovah, and T, or the cross, the Son. A writer in "Moore's Magazine" ingeniously supposes it to be a representation of three T-squares, and that it alludes to the three jewels of the three ancient grand masters. But these solutions are also suggested without any proofs, while the fact that the *tau cross* as an emblem antedates the Christian era, effectually disposes of one of them. It has also been said that it is a monogram of Hiram and Tyre, and others assert that it is only a modification of the Hebrew letter *shin,* which was one of the Jewish abbreviations of the sacred name. Oliver thinks, from its connection with the circle and triangle in the Royal Arch jewel, that it was intended to typify the sacred name as the author of eternal life.

The same objection may be made to these conjectures: no proof is advanced by their authors to support them, while the monuments and hieroglyphs of Egypt show that the *tau cross* was in use as an emblem before the era of Hiram. Dr. Mackey says that, among so many conjectures, he need not hesitate to offer one of his own, and remarks as follows:

> The prophet Ezekiel speaks of the *tau,* or the tau cross,
> as the mark distinguishing those who were to be saved,

on account of their sorrow for their sins, from those, who, as idolaters, were to be slain, it was a mark or sign of favorable distinction, and with this allusion we may, therefore, suppose the triple tau to be used in the Royal Arch degree as a mark designating and separating those who know and worship the true name of God from those who are ignorant of that august mystery.

This is much nearer the truth, but is not, after all, any explanation of either the meaning or origin of the emblem itself. It is only a suggestion of the reason why it may have been adopted by the Royal Arch degree, as being appropriate to its spirit. Dr. Mackey leaves us in the dark why Ezekiel speaks of it as an emblem of life and salvation:

> The English Royal Arch Lectures say that "by its inter-section it forms a given number of angles that may be taken in five several combinations; and, reduced, their amount in right angles will be found equal to the five Platonic bodies, which represent the four elements and the sphere of the universe."

But this, if true, throws no light on the subject. The tau cross, as an emblem in various forms, is found on the ancient monuments of Egypt, and in order to discover its real meaning, and how it came to be used as a symbol, we will have to go back to a period long before the era of King Solomon.

Q. What is the origin and meaning of the triple tau?

A. The triple tau is the ancient symbol of the tau cross, three times repeated and joined at a common center. The *tau* cross is the same in shape as the Greek letter T, which is also called *tau,* and was anciently considered as an emblem of life. It was held to be a sacred mark, and was placed upon the foreheads of those who escaped from shipwreck, battle, or other great peril of life, in token of their deliverance from death. This is why the tau is mentioned in Ezekiel (4:4-6) as the "mark set upon the foreheads of the men" who were to be preserved alive. The name by which this emblem is known points to its origin, and also the reason why it is selected as an emblem of life. The word *tau* is derived from an Egyptian or Coptic root, meaning a bull or cow, and the constellation anciently marking the vernal equinox. This word, with a Latin or Greek termination, is found in both those languages—*Taurus* (Latin), a bull, and the *Tauros* (Greek), meaning the same. The ancient hieroglyphic sign of the constellation Taurus and the vernal equinox is in the form, ♉, as an astronomical sign, representing the face and horns of a bull. It is now considered established that letters were derived from the ancient hieroglyphs, and, when the phonetic mode of writing was invented, many of those letters retained the name of the object which the original hieroglyphs, or pictures, were intended to represent.

These hieroglyphs, in process of time, assumed a form more and more arbitrary, so much so that, at last, they lost almost all resemblance to the original picture, of which, however, many of them still retained the same. It was thus that the drawing of the face and horns of a bull became a mere outline, and assuming this form ♉, as an astronomical sign. Even this did not remain permanent, for, after it came to be used as a letter, it happened, either from carelessness or convenience in writing, that the circle representing the face of the bull became a straight line. The same kind of a change appears to have taken place with the original picture of Aries, or the head and horns of a ram; which from the actual picture, became finally like this, ♈, its present form as an astronomical sign. It was in just the same way that ♉, *Taurus,* became changed, as shown by Figure 3, after it came into use as a letter. The next change was as shown by Figure 4, and, finally, the semicircle of the horns, like the circle formerly representing the face, became a straight line also, and the character assumed this form, T.

These changes may be represented at one view, as follows:

**TAURUS AND TAU CROSS**

The first of these is the original hieroglyphic picture of the head and horns of a bull; the second is the astronomical sign of *Taurus,* and, as such, for astronomical purposes has retained that form, probably because so seldom thus used in comparison to its subsequent employment as a letter; the third shows the transition of the second into the fourth, after it began to be used alphabetically, and is one form of the Greek letter tau; the last is the Greek and Roman capital tau, which is identical to the tau cross.

The common name of all these characters, it will be observed, from the first pictorial representation of the head and horns of a bull, and including the sign T, is *tau,* meaning a bull or cow. For the real name is tau, the *"us"* of the Latin and the "os" of the Greek being nothing but the usual termination characteristic of those languages. The Phoenician name of the letter T, according to Rawlinson, is also *tau,* meaning, however, *"bread"* in that language. But, as the bread is the nourisher and "staff of life," the word is equivalent to the Egyptian "giver of life." The real meaning and figurative significance of the Phoenician word for bread thus becomes at once apparent; it may have had a double as well as a figurative meaning. Even in the Egyptian the word has a meaning suggestive of agriculture and the raising of grain, out of which bread is made, for the Coptic word *thour* meant a *bull,* and its verb *athor* meant *to plow.*

The constellation Taurus was anciently at the vernal equinox, and was considered by the Egyptians (for reasons before fully explained) the emblem of a perpetual return to life; the *sign* Taurus, and consequently the tau cross, thus became the expressive symbol of the vernal equinox and of immortality. The letter, or symbol, together with the mythology connected with it, was adopted by the Greeks, perhaps, indirectly through the Phoenicians, for the Greeks claim to have been taught the letters by Cadmus, a Phoenician. The foregoing is probably the origin of the letter *tau,* and the peculiar significance attached to it.

Rawlinson, in his notes to "Herodotus," Book V, Chapter LVIII, holds that the Greeks derived their letters directly from the Phoenicians, for the reason that they are quite similar in form, and that their names all have a significance in the Phoenician language of the object which they were originally intended to represent; while, on the other hand, their names have no meaning whatever in the Greek tongue. In other words, he argues that the names of the letters are Phoenician, and not Greek, and that, therefore, the Greeks must have borrowed their letters directly from the Phoenicians. This he shows conclusively by the table of letters with their names, which he gives. This list of names, however, proves just as conclusively that the Phoenicians themselves did not invent the letters, but simply translated their names into their own language when they began to use them. The names, translated into English, are as follows:

| | | |
|---|---|---|
| A Bull, | A Paling, | A Prop, |
| A Tent, | A serpent, | An Eye, |
| A Camel, | A Hand, | A Mouth, |
| A door, | The Hollow | An Axe, |
| A Window | of a Hand, | A Head |
| A Hook, | A Prick-stick, | A Tooth, and |
| A Lance | water, | Bread |
| | A Fish, | |

The Phoenicians, it is certain, were a maritime nation. They were wholly commercial in their character, and the most renowned people of all antiquity for their naval pursuits. Had they invented the letters, the objects which the letters most certainly would have represented would have been of a marine and commercial nature. We would expect to find ships, boats, sails, ropes, rudders, anchors, chains, oars, and that class of objects. None of these, however, appear; on the contrary, the objects are all pastoral or agricultural in their character, indicative of a people engaged in those pursuits—a people who used the bull to plow with, and whose commercial enterprises were not conducted on the sea by ships.

It is another significant and almost conclusive fact that each and every one of these "objects," except the camel, are found in profusion among the hieroglyphic pictures of the Egyptians, and were in daily and familiar use in all their written inscriptions, as we find them on their monuments and sculptures even to this day. This is true of no other ancient people, and the conclusion becomes irresistible that the Phoenicians, whose ships and traffic brought them in frequent contact with the Egyptians, borrowed of them their letters or derived them from the hieroglyphics of Egypt. They naturally, and almost of course translated the names of the various objects and animals represented in the hieroglyphs into their own Phoenician tongue. This the Greeks, when they in turn borrowed from the Phoenicians, did not do, probably because, when the hieroglyphs reached them, they had assumed a more arbitrary form, and one so far removed from the original pictures as to render any such translation wholly unnecessary, if not impossible.

That the Phoenicians, a people preeminent for their ingenuity and skill, greatly improved on the Egyptian method, and reduced the hieroglyphs to a more strictly alphabetic and arbitrary form and use, is highly probable, if not certain; but that the originals of the letters, together with their names, first came from Egypt, is also just as certain. The improvements which the Phoenicians made in the art of writing by letters was, no doubt, as much due to the fact that they were free from certain religious obligations, which hampered an advance in that

direction by the Egyptians, as to their own characteristic ingenuity and national aptitude for scientific pursuits.

It may be urged, as an objection to our derivation of the
letter tau, that, in the Phoenician and Hebrew alphabets, the
letter A is named *aleph,* meaning a bull. The Greeks, also,
called the letter A *alpha,* adopting the Phoenician name. But
the *sound* of A is also represented in the Egyptian hieroglyphs
by the tau cross. The very fact, therefore, that the Phoenician
letter A was named a bull, shows that the Egyptian tau cross
had a name with a similar meaning, and *did* represent not
only a bull, but specifically the sacred bull called *Apis,* which,
according to the Egyptian system, gave it the sound of the letter A, for the use of the hieroglyph as a letter followed the first
sound of the name of the object represented. It also shows
that the allusion of the tau cross of Egypt was to the vernal
equinox, and the constellation of the bull thereon, for which
reason it was an emblem of life and a return to life. Apis was
the name of the sacred bull, under which emblematic form the
Egyptians worshipped Osiris, the sun-god.

In the Chaldaic alphabets it is the letter T which is said to
have been originally represented by a bull In the alphabet of
Cadmus the letter T is a cross, similar to another of the Egyptian signs for the letter A. Now, if all these alphabets were in
fact originally derived from the hieroglyphics of Egypt, this is
just the sort of confusion which we would naturally expect to
exist respecting the name and form of the letters T and A
among the earlier alphabets of other nations, who translated
the names into their own language, and began to use them on
the Egyptian system, and according to the initial sounds of
those names.

In some of these alphabets the letter A, while it lost the
form of the cross, retained the name of a bull, as no distinction
would naturally be made by other nations between that particular bull named *Apis,* sacred to the Egyptians only, and a bull
generally.

In other alphabets both the name and form might be retained, but the name being translated into another language, the letter might be used as the symbol of another sound. The Greek *Tauros* and Latin *Taurus* have the word *tau* as a common root, which may have been derived from the Egyptian or Coptic *kau,* a cow or bull, or *athor* The Arabic *thour,* a bull is evidently the same as *athor,* the "a" only being dropped. Such changes as these would cause the hieroglyphic sign of the bull to represent in some languages the sound of T in place of that of A.

**CRUX ANSATA, EMBLEM OF ETERNAL LIFE** (EGYPTIAN)

The specific ancient Egyptian *"emblem of eternal life,"* however, does not appear to have been adopted in its complete form by other nations—that is, as a letter. Its form was abbreviated, although its symbolical meaning was retained to some extent. The Egyptian symbol of eternal life, in its unabridged form, is as below, and was known in later times as the *"Cruz Ansata." As* will be *seen,* it is nothing more than the "tau cross" surmounted by a circle, sometimes made somewhat oval in shape. The entire hieroglyphic was probably originally the picture of the head and horns of a bull, *surmounted by the orb of the sun,* thus expressing in a still more direct and specific manner the sun in Taurus.

It was thus they were accustomed to represent *Apis.* This symbol, from its constant use at first as a sacred emblem, and finally, as a letter, or hieroglyphic, would naturally assume more and more of an arbitrary form. The face and horns of the bull would gradually take the shape of a cross, as before described, and the orb of the sun which surmounted it lose somewhat its perfect circular form. The whole hieroglyph would thus finally assume an arbitrary form, like that here represented. If this conjecture be correct, it fully explains why this peculiar symbol denoted among the Egyptians eternal life— the reason for which, according to both Wilkinson and Kendrick, has as yet remained in obscurity. (See Kendrick's "Ancient Egypt," vol. i, page 254; Wilkinson's "Ancient Egyptians," vol. 1, page 277.)

**GODDESS OF TRUTH
AND JUSTICE HOLDING
CRUX ANSATA**
(WILKINSON)

This Egyptian emblem was subsequently named the *Crux Ansata,* or "cross with a handle," because it was thought the circle was nothing more than a handle for the purpose of carrying the cross. It is, in fact, often represented as being so carried on the sculptures, but quite as frequently otherwise. The following cut shows the "sign of life" held by the lower end, in the hand of the double goddess of Truth and Justice.

The idea advanced by some, that it is a key, derives little or no support from the monuments; besides this, the Egyptian form of a key was entirely different, as is seen from the drawing which accompanies our explanation of the masonic emblem of "the key."

The *Crux Ansata* was adopted by the early Christians of the East as an appropriate symbol of their faith. The old

inscriptions of the Christians at the Great Oasis are
headed by this symbol, and it is also found in some of
their monuments at Rome.                    (Wilkinson)

Among the ancients the cross in this form, **+**, was also considered a sacred emblem, as it pointed to the four quarters of the
heavens, and embraced both the celestial and terrestrial hemi-
spheres. It was thus a symbol of the universe, and expressive
of the perpetual life and endless duration of nature. The Rosi-
crucians also taught that this form of the cross was the symbol
of *light,* because it contained in its formation the ancient
Roman letters LVX, *lux,* the Latin word for *light.* Whether this
beautiful conceit was invented by them or derived from
ancient sources is unknown.

The tau cross, is as has been shown, an ancient symbol of
Egypt denoting salvation and eternal life. The triple tau, being
a combination of the tau cross *three times* repeated, teaches us
that "we have an immortal part within us that shall survive the
grave, and which shall never, *never,* NEVER die" (Masonic
Lecture).

## The Astronomical Triple Tau

Q. Has the triple tau any further astronomical signification?

A. It has—for, when the geometrical principles upon which it
is erected are analyzed, it will be found to represent, symbolically, the Royal Arch, together with its three principle points,
and many other astronomical particulars. In order to explain
this more fully, let us draw out on our "trestle-board" a triple-
tau. We will first draw the line *A B* (see the following
diagram), representing the great equinoctial colure; on this
describe a semicircle, and erect the Royal Arch (see illustration
on page 69). Next distinguish the two equinoctial points by
two parallel lines, in the same manner as the solstitial points
are marked in the emblem of "a point within a circle" (see
page 130). Draw the line *C D,* representing the summer solstice and tropic of cancer, in the same manner as shown in the

**GEOMETRICAL
TRIPLE TAU**

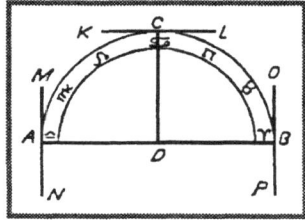

emblem last referred to. The lines at the extremities of *A B* are perpendicular to it, and in a properly drawn triple tau the lines *M N, O P,* and *K L,* are all equal to each other, and equal to the radius of the circle which may be inscribed within them. The radius of any circle is one sixth of its circumference, and, therefore, is a *chord* of an *arc* of 60°. It therefore follows that the line *K L* is divided by the perpendicular C D into two parts, each of which represents 30°, or one sign of the zodiac. The same is true of the lines M N and *O P,* each of which is divided by *A B* into parts representing a *chord* of 30°. The line *O P* is thus the *chord* of the two signs ♈ and ê, the line *K L* is the *chord* of ♊ and ♋, and the line *M N* of ♌ and ♍, which constitutes 180°, and takes us to the first point of Libra (♎), at the autumnal equinox.

The first six signs of the zodiac, reaching from the vernal to the autumnal equinox, and constituting the Royal Arch of heaven, are therefore represented with geometrical precision by the exterior lines of the triple tau, while, at the same time,

**ASTRONOMICAL
TRIPLE TAU**

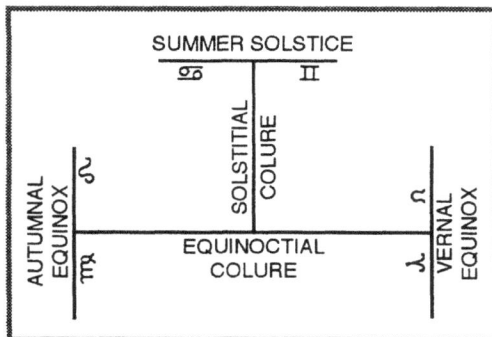

the line *K L* represents the summer solstice, and the lines *O P* and M N the vernal and autumnal equinoxes. This geometrical analysis of the triple tau reveals the fact that it is a striking symbol of the Royal Arch, and the exaltation of the sun therein, and several other astronomical particulars. This drawing is what may be termed the astronomical triple tau.

The three principal points of the zodiacal arch as explained on page 69, are emblematic of wisdom, strength, and beauty; on these the whole arch of heaven seems to rest. The three parallel perpendicular lines, as they represent those three points, are also emblematic of wisdom, strength, and beauty, and, as a perpendicular line is the geometrical symbol of a pillar, they may be said to denote the three great masonic columns placed in a triangular form. It was these emblematic pillars that Job alluded to when, speaking of T. G. A. O. T. U., he said, "The pillars of heaven tremble and are astonished at his reproof (Job 26:11). The three masonic columns of "wisdom, strength, and beauty," must not be confounded with the "pillars of the porch." The latter have a different emblematic meaning, which has already been explained.

### The Quadruple Tau

That part of the zodiac embracing the summer solstice, and reaching from the vernal to the autumnal equinox, was con-

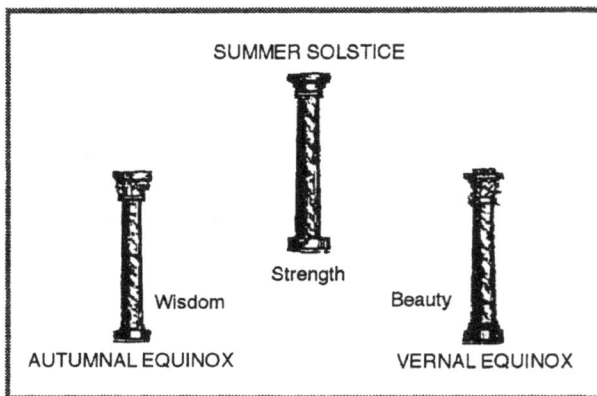

PILLARS OF WISDOM, STRENGTH, AND BEAUTY

SUMMER SOLSTICE · Strength · Wisdom · Beauty · AUTUMNAL EQUINOX · VERNAL EQUINOX

sidered the most important and sacred by the ancients, because the sun was therein exalted, and because it embraced the whole of the seasons of springtime and harvest. It is that part of the zodiac only which is therefore represented in the symbol of the triple tau. If, however, we unite in one emblem

**TRIPLE TAU AND CIRCLE EMBORDERED WITH PARALLEL LINES COMBINED**

the triple tau and that of "a circle embordered by two parallel lines," we will have a correct geometrical representation of the whole zodiac, the four principal points of which will also be designated in a similar manner, by which it will be seen that the two emblems are in fact but parts of one complete whole.

The union of these two masonic emblems gives us the device which appears between them in the above diagram, which, as will be seen, is another ancient and well-known emblem, sometimes called the "cross of Jerusalem." It consists of the tau cross four times repeated, and joined at a common center, which is really that of the zodiac. The circle about that center is sometimes exhibited in this emblem, but is more frequently left out, as not being required to express its meaning, and adding nothing to its beauty. This emblem would be more properly known under the name of the *quadruple tau.* This emblem was brought by the Crusaders from the East, and they, ignorant of its true meaning, adopted it as the symbol of their faith, from its supposed resemblance to the Christian cross.

The quadruple tau represents at one view the entire universe. The central lines, one of which is *horizontal* and the other *perpendicular,* thus crossing each other at *right angles,* point to and embrace the four quarters of the celestial and terrestrial spheres. The limits of the sun's circuit amon the stars, both at the solstitial and equinoctial points, are designated by

QUADRUPLE
TAU

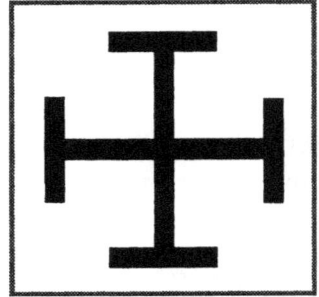

the lines at the extremities of the central one, placed at right angles to bar the way. Two of them represent the solstitial points, which is in entire harmony with the emblem of the "circle embordered by two parallel lines," from which they are derived, as explained in our description of that emblem on page 129. The other two, taken from the triple tau, represent the vernal and autumnal equinoxes, as has been explained in our remarks on the astronomical triple tau. The quadruple tau, moreover, being composed entirely of "right angles, horizontals, and perpendiculars," contains within itself all the secret signs of Freemasonry, a fact which I am not permitted to further explain. It will, however, be apparent to every "bright mason," who can soon study them all out for himself.

Q. Are there any remaining masonic emblems which have not been explained?

A. The *gavel,* the rough and perfect *ashlar,* the *twenty-four inch gauge,* the *trowel,* the *plumbline* and *level;* also the *mallet, chisel,* and pickaxe appertaining to the Royal Arch degree, have not been astronomically explained, because all of them are nothing more than the mechanical tools of those operative masons and architects who (as will be subsequently explained), after the Mysteries ceased to be celebrated, assumed entire control of our Order, and which they ingrafted into the ancient ritual at a comparatively recent date, as emblems of their art. Had they at that time *invented* the whole ritual, originated the entire matter, no other emblems but those of a like exclusive mechanical import would have been adopted. Those other sublime astronomical allegories and pure scientific symbols, with the elevated philosophy they teach, would never have been found in Freemasonry. We are indebted to Preston, Webb, and Cross for a fine moral application of the gavel, twenty-four inch gauge, etc. They require no further explanation.

With the exception of these, all the other ancient symbols have been found to have an astronomical origin and meaning. It is also a strong confirmation that no contradiction exists among them when thus explained. The *separate* astronomical explanation of each *one* of them is in perfect harmony, not only with all the others, but also with the main central allegory of the annual passage of the sun among the stars of the zodiac, his death during the winter months, his return to life at the vernal equinox, and his exaltation at the summer solstice.

## The Words "Mystery" and "Masonry"

Q. Is there any connection between the words "mystery" and "masonry"?

A. If, in fact, the masonic institution, as Mackey and Oliver both admit, was descended from the ancient "Mysteries," there should be some close connection between the words "mystery" and "masonry," even if the latter is not directly derived from the former. The word "mystery," which originally had an exclusive meaning, came in process of time to have three different meanings, all derived from the original one:

1.   It was the name of the *sacred drama* which constituted the ceremony of initiation into the secret religious associations of the ancients, which were so named from the fact that the "aspirant" for initiation was blindfolded. The word "mystery" is derived from the Latin *mysterium,* from the Greek μυστήριον, from μυστής, from μυεὶν, *to shut the eyes.*

2.   In the middle ages it came to be applied to a different sort of "sacred drama," founded on the legends of the Chris-

tian religion. These "Mysteries" or religious dramas, were, however, performed in public, and had no element of secrecy about them.

3. Another use of the word "mystery" arose from the fact that all scientific knowledge was formerly concealed in the ancient Mysteries, and communicated only to the initiated. Great skill, therefore in any art which required scientific knowledge, anciently implied initiation into the Mysteries. Hence, in process of time, and even after the Mysteries themselves were suppressed, the word "mystery" was applied to any art which required scientific knowledge in addition to manual dexterity. The art of architecture is one which requires not only a proficiency in geometry, but several other sciences. In more ancient times, owing to the peculiar position and construction of temples, considerable knowledge of astronomy, even, was required by the architect. This art was therefore preeminently above all others denominated a "mystery," and the words "mystery" and "masonry"—i.e., architecture—became synonymous in meaning. Architecture was thus probably the first one of the arts called a "mystery"; this name, however, at length came to be applied to all the arts without distinction, including even those wholly mechanical.

There can be no doubt that all the early architects, at least, like the Tyrian artists who directed the work at the building of King Solomon's temple, derived the scientific knowledge required for their profession from having been initiated into the Mysteries of Dionysus. The word "masonry" has been thought to be derived from several different roots, by different writers, but it is not so far removed either in form or meaning from the word "mystery" but that it might not have been derived either directly or indirectly from it. In fact, Hutchinson, in his "Spirit of Masonry," advances the idea that the word is derived from a corruption of the Latin *mysterium,* but fails to give any satisfactory reason for his opinion. The foregoing considerations, however, tend to show that his *conjecture* is not without some support. The derivation of the word "mason" from the french *"magon,"* a house, will only take us back to the Mysteries by another path, for the word *"magon"* is derived from the Latin *maceria,* a wall or inclosure, which carries with it the idea of secrecy, and the exclusion of all who have not a right to enter. Thus, all those who were not initiated into the Mysteries were called the profane—i.e., *pro-fano,* those without the temple—and who had no right to enter at all times. The words "temple" and "house" were also anciently synonymous. (See 2 Kings 6:7-9; also, 2 Chron. 3.) Brother J. H. Little, formerly G. H. P. of Virginia, derives the word "Freemasonry" directly from the Egypto-Coptic, and uses the following language on the subject:

> Great mistake has arisen from the very name we bear, and many do not understand what we are, or what our name itself means. Masons are not free, in the sense in which the word is sometimes use; they are positively bound by absolute laws, they are the slaves of truth and their word—unqualified obedience to their duty. The profane are free, the mason is not. The origin of our name shows this. Our title is "Freemason," and this is not an English word, nor is our Order of English origin. The name is not of any of the languages of modern Europe, nor is it found in the classic tongues of Greece and Rome; nor is it a part of the languages of Syria, Tyre, or Chaldea, nor is it Hebrew. More ancient than all, it comes from a nation that had organization, architecture, and literature, before Abraham first beheld the stars glitter above the plains of Shinar. It is from the language of ancient Egypt; that wonderful land where all antediluvian science and art was preserved and extended, where a system of priestly and kingly government was carried out which has been the wonder of the world; that land where men of science, organized into a close and secret organization, ruled; where they created a mystic language, and where they erected those mighty works of architectural skill whose undestroyed firmness still amazes the world—among these ancient

sages the sun was an object of veneration, as the visible power of life and light. In their language it is called *Phre,* and in the same language *mas* means *a child.* Hence, being born of light, that is, knowledge of every kind, physical, moral, and intellectual, they called themselves *Phre-massen*—Children of the Sun, or Sons of Light. They inculcated and practiced purity and perfection of the body, control of all the passions, or moral purity, and devoted themselves go the intense study of all intellectual acquirements. Now, this is Freemasonry—we are true Sons of Light.

(St. Louis "Freemason's Monthly," January, 1872)

Q. How came operative architects, or masons, to be the last custodians of the secrets of the ancient Mysteries?

A. It has no doubt been a puzzle to more than one, why the architects and temple-builders of antiquity should have been so intimately connected with the Mysteries, and thus have been in a position to hand down their essential secrets and philosophical teachings, from generation to generation, to those skilled workmen who came after them. In other words, how was it that the *operative* masons, or architects, became special guardians, and their guilds, or associations the depositories of these philosophical mysteries? If a good and sufficient answer to this question can be found, one great stumbling block and source of skepticism will be removed. This question we think we can answer. The ancient Mysteries, as is well known, were celebrated in the hidden recesses of the temples. In order to present the grand and impressive drama of initiation, many secret chambers, doors, and labyrinthian passages had to be constructed within the interior; also, much ingenious mechanism, by which wonderful and sublime spectacular effects were produced. It was, therefore, a matter of necessity that the building of a temple (except the bare outside walls) should be intrusted only to those who had been duly initiated. Any "tattling mechanic" might otherwise disclose the whole secret. Such operative architects and artists, therefore, who were known and distinguished as the most cunning workmen, were initiated in all branches of the Mysteries, because their services were imperatively necessary.

Among the buildings uncovered at Pompeii is a temple of Isis, which is a telltale of the Mysteries of the Egyptian deity, for the secret stairs which conducted the priests unseen to an opening back of the statue of the goddess, through whose marble lips pretended oracles were given and warnings uttered, now lies open to the day, and reveals the whole imposition. ("A Day in Pompeii," "Harper's Magazine," vol. ii.)

> When the sages of India conducted Apollonius to the temple of their god, singing hymns and forming a sacred march, the earth, which they struck with their staves in cadence, was agitated like a boisterous sea, and raised up nearly two feet, then calmed itself and resumed its usual level. The act of striking with their sticks betrays the necessity of warning workmen, who were placed beneath, to raise a moving stage covered with earth—an operation plainly effected by the aid of mechanism, very easy to be comprehended. It is probable a similar secret existed in other temples. English travelers who visited the remains of the temple of Ceres, at Eleusis, observed that the pavement of the sanctuary is rough and unpolished, and much lower than that of the adjacent portico. It *is* therefore probable that a wooden floor on a level with the portico covered the present floor, and concealed a vault designed to admit of the action of machinery beneath the sanctuary for moving the floor. In the soil of an interior vestibule they observed two deeply indented grooves, or ruts, and as no carriage could possibly be drawn into this place, the travelers conjectured that these were grooves to receive the pulleys which served in the Mysteries to raise a heavy body—"perhaps," said they, "a moving floor." In confirmation of this opinion, they perceived further on other grooves which might have served for the counterbalances to raise the floor; and they also detected places for wedges, to fix it immovable at the desired height. These were eight holes fixed in blocks

of marble, and raised above the floor, four on the right and four on the left, adapted to receive pegs of large dimensions.

We are also informed that, in order to descend into the caves of Trophonius, those who came to consult the oracle placed themselves before an aperture apparently too narrow to admit a middle sized man; yet, as soon as the knees had entered it, the whole body was rapidly drawn in by some invisible power. The mechanism used for this purpose was connected with other machinery, which at the same time enlarged the entrance to the grotto. The person who went to consult this oracle was obliged to make certain sacrifices, to bathe in certain rivers, and to anoint his body with oil. He was then clothed in a linen robe, and, with a cake of honey in his hand, he descended into the grotto in the manner before described. What passed there was never revealed, but the person on his return generally looked pale and dejected. The individual whose name this cave bore was an architect of great skill, and in conjunction with his brother, Agamides, was the architect of the temple of Apollo, at Delphi; and they were, of course, the designers and constructors of all the mechanical secrets of that temple, no doubt far more ingenious and terrifying in their nature than those of the oracular cave just described. The Mysteries being also celebrated in the temple, the demand for secrecy was imperative, and the priests, fearing that the initiation of Trophonius and Agamides would not insure their silence, resorted to assassination. The brothers were desired by the god, through the priests, to be cheerful, and to wait eight days for their reward; at the end of which period they were found dead in their beds—the result of poison, or some other secret means of murder. (See Salverti's "Philosophy of Magic," vol. 1, Chapter XI).

Instances might be multiplied of the secrets involved in the construction of ancient temples, which made it a matter of necessity that the architects should be initiated, if allowed to live. But enough has been advanced to make it plain that the initiation of operative architects was a matter of absolute necessity, When the Mysteries were discontinued, after the advent of Christianity as the state religion of the Roman Empire, it was no longer necessary for the temple-builders to belong to any such organization, but by that time operative architects had found that the bond of union which the initiation into the Mysteries had established among them was useful and profitable. It enabled them to keep the higher secrets of their art among themselves, thus giving them a monopoly of the whole business of temple-building. They were thus also enabled to assume an independence and consequence, upon which followed the favor of princes and those high in authority, who desired their services to erect a palace or build a cathedral. The operative architects, therefore, kept up their secret organization, and thus preserved the occult tie which originally united them in the Mysteries, of whose legends, signs, and emblems they became the last custodians, after the Mysteries themselves had fallen into disuse, and ceased to be celebrated either at Athens or Rome. Thus originated those mysterious "travelling Freemasons" of the middle ages, who left so many "massive monuments of their skill" as early as the ninth and tenth centuries. Thus, also, originated those famous guilds of operative masons of the fifteenth, sixteenth, and seventeenth centuries. No other hypothesis will satisfactorily account for the strange character and mysterious nature of those secret associations of operative architects.

Although the Mysteries themselves are traced back historically to the days of ancient Egypt, yet there is no chronological impossibility, or even improbability, of their connection with the societies above mentioned, for they were celebrated in some form as late as the eighth and perhaps twelfth century, while the traveling Freemasons are traced back to the eighth or tenth century. Notwithstanding the celebration of the Mysteries was prohibited by the Christian emperors succeeding

Constantine, as being connected with the pagan worship, yet many of their rites continued to be observed under assumed names and the pretense of *convivial* meetings, for a long time afterward (Gibbon, Chapter XXVIII). Maximus, Bishop of Turin, writes in the middle of the fifth century against the ancient worship, and speaks of it as if existing in full force in the neighborhood of his city. The Eleusinian Mysteries at Athens, indeed, seem to have enjoyed a special exemption, for Gibbon informs us that the Emperor

> Valentinian immediately admitted the petition of Praetextatus, proconsul of Achaia, who represented that the life of the Greeks would become dreary and comfortless if they were deprived of the invaluable blessing of the Eleusinian Mysteries.

This petition was, no doubt, accompanied with an assurance that the secret doctrines taught in the Mysteries, being those of the unity and spiritual nature of God, and the immortality of the soul, were not inconsistent, but rather in harmony, with the Christian religion, which would account for the petition being so promptly granted. The Mysteries at Athens, in consequence, although suspended, do not seem to have ever been totally suppressed, but continued to be celebrated in some form as late as the eighth century. It is also certain that the Mysteries, under various forms, continued to be celebrated in Britain and on the Continent as late as the tenth century. Dr. Oliver says, in his "History of Initiation,"

> We are assured, on undoubted authority, namely, from the bardic writings of that period, that they were celebrated in Wales and Scotland down to the twelfth century of Christianity.

This brings us down to an era when it is admitted on all hands that the travelling Freemasons existed, by whom, some claim, our fraternity was invented. It is not, however, claimed that the Mysteries in their purity or original splendor existed at so late a period. No doubt they had become corrupt, and many of their secrets had been lost. No doubt they had become obscure, but still they existed, impressed with their original character. The connection is therefore close between them and the mysterious secret rites and ceremonies of those societies of operative masons and architects above mentioned. When, in process of time, the celebration of the ancient Mysteries in a modified form was confined to these associations of operative architects, for the reasons before given, then it was that the term *Freemason* began to be descriptive of the *initiated.* This would more rationally account for the present name of our fraternity than the ingenious derivation of the words "free-mason" from the Egyptian roots, *Phre-massen* (Children of Light) as advanced by Brother J. H. Little.

Salverti, in his "Philosophy of Magic," is of the opinion that the occult sciences, possessed by the secret societies of the middle ages in Europe were derived from the learning taught in the Mysteries. He says:

> It is certain that, in that age of ignorance, learned men have conveyed the charge of their knowledge to secret societies, which have existed almost in our day. One of the brightest geniuses who shed honor upon Europe and the human race, Leibnitz, penetrated into one of these societies at Nuremberg, and, from the avowal of his panegyrist [Fontenelle, "Eloge de Leibnitz"], obtained there instructions which, perhaps, he might have sought for in vain elsewhere. Were these mysterious reunions the remains of the ancient initiations? Everything conduces to the belief that they were, not only the ordeal and the examination, to which it was necessary to submit before obtaining an entrance to them, but, above all, the nature of the secrets they possessed, and the means they appear to have employed to preserve them. (See "Philosophy of Magic," vol. 1, Chapter XI)

But if, as Salverti learnedly argues, the scientific secrets of the Mysteries were thus transmitted to the secret societies of the middle ages, we may be certain that not only the form of initiation in substance, but also many of the legends or scientific allegories, as well as the symbols and emblems connected therewith, were also handed down in like manner, and the same may probably be said of many of the signs and modes of recognition. In this connection it is worthy of remark that none of the passwords of Freemasonry are either English, German, or French, nor indeed of any modern spoken language. Had Freemasonry been invented, or fabricated, either in Germany, England, or France, such would not have been the case. We might as well expect to find the armies of France, Germany, England, or America, using Coptic, Chaldean, and Hebrew countersigns, as the Freemasons do, had our fraternity originated in either England, France or Germany.

## The Antiquity of Masonry

Q. What is the probable antiquity of masonry?

A. There can be but little doubt that the Mysteries, from which, as we have seen, Freemasonry is the direct descendant, were first arranged when *Taurus* was on the vernal equinox, *Leo* at the summer solstice, and *Scorpio* at the autumnal equinox. The solar allegory, as handed down to us, shows this to be the fact. At the rate of the precession of the equinoxes is known, we can calculate when the vernal equinox was in *Taurus.* Such a calculation will take us back about four thousand two hundred and eighty years. The antiquity of masoniy is thus written on the face of the starry heavens—a record which utters no falsehoods.

## Freemasonry Not Sun-Worship

Q. *Is* it to be understood, from the foregoing pages, that Freemasonry is nothing more than a fragment of an idolatrous form of sun-worship?

A. Such is far from being the case, nor has anything been advanced in the foregoing pages which, unless wholly misunderstood, gives any countenance to such an idea. In the introductory chapter it was fully shown that the Mysteries themselves, in their primitive and uncorrupted form, taught the unity of God and the immortality of man as their cardinal doctrines, and that the sun was but a *symbol* of him whom "the sun, moon, and stars obey, and beneath whose all-seeing eye even comets perform their stupendous revolutions" (Masonic Lecture).

Though in all parts of our ritual, from the threshold to the altar, and from the altar to the *penetralia* (as in the ancient Mysteries, from which Freemasonry has descended) the profoundest truths of science and true religion are taught and illustrated by astronomical allegories, yet nowhere do we find, even in its most ancient portions, any prayers, invocations, or adoration, addressed to the heavenly bodies themselves. The sun and the hosts of heaven are only used as emblems of the Deity—a sacred symbolism, with which the Bible itself abounds.

In more ancient times, when false and idolatrous forms of religion ruled all the civilized nations, masonry protected the worshippers of the true God. This was not only true in Rome and in Greece, where Socrates and Pythagoras fell martyrs to truth, but also in Palestine. When we call to mind the long succession of Hebrew kings "who did evil in the sight of the Lord," and sacrificed to Baal "upon the high places and in the grove," a crime of which even Solomon was guilty in his old age, we can easily see that, except at certain favorable epochs, it was not safe, "no, not even in Judea," to deny the actual divinity of the sun, moon, and stars. The Jews stoned the prophets just as the Greeks persecuted the philosophers. The great debt that not only religion but science owes to masonry can hardly be estimated.

In its ritual, as we have *seen,* most of the truths of astronomy and geometry are illustrated and perpetuated. And it would be no stretch of the imagination to say that, if all, whether of books or manuscripts, were swept out of existence, the ritual of our Order, as orally communicated, would alone be sufficient to transmit to future generations a knowledge of the true God and a correct code of morals, as well as the leading principles of science, whereon to build anew the great temple of knowledge.

# Chapter 9

# *CONCLUSION*

THIS WORK might with perfect propriety have been named "A Defense of Freemasonry"—

1. Against all the assaults of those who stigmatize all its claims to a remote origin as delusive and false. This class of objectors assert that the Order is of no great antiquity, having originated late in the middle ages, in a union of operative stonemasons, builders, and carpenters, who thus sought to keep secret the practical arts of their craft, and also by a cooperative combination to be able to control the business of architecture, and fix the rate of wages for skilled workmen, on the same principles of the "trade-unions" of the present day. Such organizations, without doubt, did exist, but they could never have originated the profound, beautiful, and scientific *astronomical* allegory of the masonic legend. This has already been made evident to the reader, without argument.

2. Against the absurd claims of a class of over-enthusiastic masonic writers, who, going to the opposite

extreme, affirm that masonry originated in the garden of Eden, by inspiration of God; that Adam was the first Grand Master, he being succeeded by Enoch, Shem, Ham, and Japheth, Abraham, Moses, Solomon, and so on down to General Warren, who fell at the battle of Bunker's Hill! These well-meaning enthusiasts, provoked by a lively imagination, see masonry in everything, and claim that every structure ever built, from the Tower of Babel and the pyramids to King Solomon's temple, the Colosseum at Rome and St. Paul's Church in London, were built by the selfsame Order which now assembles in its lodges in Europe and America, Asia and Africa, under the name of Freemasons. These absurd claims only serve to bring masonry into ridicule, and cause judicious persons to laugh at our supposed credulity, thus doing the fraternity more real harm than the former class, who really accord us a very respectable age of eight or ten centuries.

One great stumbling-block in the way of rational investigation is caused by extravagant expectations, and an unphilosophical demand for a too *exact* correspondence between alleged ancient masonic organizations, and the emblems relating to them, with modern masonic bodies, their degrees, emblems, verbal rituals, and the modern version of our ancient legends. Many worthy brothers, among whom are some of much learning, seem to entertain the idea that unless we go to the full extend of demonstrating that the ancient Mysteries were identical in all respects with modern masoniy, including not only our present ritual and lodge-work, but also the division and order of the degrees, that our arguments amount to nothing, and afford no proof of the antiquity of our fraternity.

Nothing less, I fear, would convince this class of investigators than the discovery of the whole ritual and catechism, beginning at "From whence came you?" etc.—as authorized by the Grand Lodge of their State—sculptured in hieroglyphics, or written in Coptic on a roll of papyrus from an Egyptian tomb. Certainly all such expectations are unreasonable and unphilosophical. Rest assured we will never find any proof that *lodges* exactly like ours, presided over by a *Worshipful Master* and *Wardens,* and conferring the *Entered Apprentice Fellow-craft,* and *Master's* degree, existed in ancient Egypt.

Freemasonry, like the Christian system of theology and mode of worship, has undergone many modifications since the day of its advent; yet, like Christianity, it has preserved its *identity,* as well as all of its vital principles and most exalted features in all ages. Although in masonry there has been no "apostolic succession," beginning with the Grand Hierophant of the Osirian Mysteries and ending with the present highly respected Grand Master of New York, yet the identity of our Order can be traced from a remote antiquity just as satisfactorily from a remote antiquity just as satisfactorily as the identity of the Christian religion can be traced from our Protestant churches (who deny all "apostolic succession") on beyond the Reformation, and through the Catholic Church, with, in earlier times, its half-pagan rites, back to the plains of Judea and the advent of Christ. Nor does the vast difference which such a view of Christianity discloses, in doctrine, practice, ritual, and mode of worship at different eras in the past, or at present in different lands and among different sects, at all obscure the real identity of the Christian system in all ages since its promulgation.

In like manner the antiquity of our fraternity and its identity are established—not so much by any such close correspondence of our present ritual and emblems with those of ancient times (as some investigators illogically look for), as from other considerations. It is quite enough if we are able to discover in ancient times, when *polytheism* was the dominant state religion in all nations, societies possessed of *similar organizations,* and, like Freemasonry, teaching the two great doctrines of the unity of God, as ONE ETERNAL *Spiritual Being,* and the immortality of the soul of man—societies like masonry, *secret in their nature, and possessed of words, signs, and other occult modes of recognition,* also of similar but not identical form of initiation, the ceremonies of which were founded upon a *similar legend, allegory or myth, the same in substance,* and only differing as to the name, era, and nationality of its hero—societies which taught the *same truths* by similar and in many cases the *very same emblems, signs, and symbols.*

These things certainly demonstrate the identity of modern Freemasonry with those ancient organizations, just as conclusively as the identity of modern Christianity, as a system of religion, with that of the first century or any intermediate time, is established by a like train of reasoning and correspondences. If, on the contrary, we confine our attention to the present condition of Freemasonry, as disclosed in the various degrees and "rites" into which it has divided itself, just as Christianity has split into Catholics and Protestants, and the latter again into numerous sects—if we regard nothing but the *verbal* form of our ritual—it is easy to show that masonry is not of any very great antiquity. The date and even the authorship of some parts of our *verbal* ritual can be and have been traced, but neither the Chevalier Ramsey nor yet those who met at the famous "Appletree Tavern," in 1717, were the founders and inventors of Freemasonry, any more than Luther and Wesley were the authors of the Christian religion.

If we view masonry from a rational standpoint, and contemplate its mystic legends and allegories in their substance, without regard to the modern language in which they are now clothed; if we investigate the meaning of its ceremonies, without regard to the specific words now used in conducting them; if we study the signs, symbols, and emblems, disregarding the erroneous modern explanation given to many of them—the great antiquity of masonry is at once apparent. It is now admitted on all sides that all the ancient Mysteries were identical, and had a common origin from those of Egypt, a conclusion which has been reached by the same method of reasoning comparison. The legend of Osiris is the parent stock from which all the others came, but in Greece and Asia Minor the name of Osiris disappeared, and those of Dionysus and Bacchus were substituted, while in the Hebrew-Tyrian temple legend the name of Hiram is found. The claim, however, that the legend of Hiram is an actual history, descriptive of events which really took place about the time of the building of King Solomon's temple, must be abandoned by the few who still blindly cling to it.

Masonry can no longer hope to stand without criticism in this age of inquiry. There is a spirit abroad which does not hesitate to catch Antiquity by its gray beard, state into its wrinkled face, and demand upon what authority, of right reason, or authentic history, it founds its pretensions. The masonic tradition cannot hope to escape examination in its turn; and, when it is examined, it will not stand the test as claiming to be *historically* true. If, then, we have no explanation to offer, it must be discarded, and take its place among many other exploded legends of the past. By showing, however, that it is not *intended* as an actual history, but is really a sublime allegory of great antiquity, teaching the profoundest truths of astronomy, and inculcating, by an ancient system of types, symbols, and emblems, an exalted code of morals, we at once reply to and disarm all that kind of criticism. The masonic Order is thus placed on a loftier plane, and assumes a position which challenges the respect and admiration of both the learned and virtuous; the learned, because they will thus be enabled to recognize it as the depository of an ancient system of scientific knowledge; the virtuous, because the Order also stands revealed to them as having been in past ages the preserver of true worship, and the teacher of morality and brotherly love. It has been the boast of masonry that its ritual contained great scientific as well as moral truths. While this was plainly the fact as to the moral teachings of our Order, to a large number of our most intelligent brothers the key which along could unlock the masonic treasury of *scientific* truth appeared to have been lost. We believe that key is at length restored; for, if the masonic traditions and legends, with the ritual illustrating them, are regarded as astronomical allegories, the light of scientific truth is at once seen to illuminate and permeate every part. If the explanation given in the foregoing pages is correct, any person who fully understands the meaning and intention of the legends and ceremonies, symbols, and emblems of our Order, is necessarily well informed as to the sciences of astronomy and geometry, which form the foundation of all the others.

And why is not the explanation correct? Have you ever considered the "calculus of probabilities," as applied to a subject like this? That masonry should contain a single allusion to the sun, might *happen,* and imply nothing. The same might be said if it contained but three or four; but when we find that the name of the Order, the form, dimensions, lights, ornaments, and furniture of its lodges, and all the emblems, symbols, ceremonies, words, and signs, without exception, allude to the annual circuit of the sun—that astronomical ideas and solar symbols are interwoven into the very texture of the whole institution, and, what is still more significant, that there is such a *harmony of relation* existing between *all these astronomical allusions* as to render the *whole ritual* capable of a perfect and natural interpretation as an astronomical allegory, *which is also one and complete*—the probability that it was *originally so intended* is overwhelming, and amounts to a positive demonstration. There are millions of probabilities to one against the theory of the allegory being accidental and not designed.

Can any reasonable mind suppose that, when Bunyon wrote his "Pilgrim's Progress," the story was an allegory of the trials and triumphs of a Christian life by an *accident* only, and that the author if it never intended or designed the allegory at all? Yet the astronomical allegory of the masonic legend pervades all parts of it, and is just as complete and perfect when examined as the allegory of the travels, combats, adventures, and temptations of the hero of "Pilgrim's Progress." The probability that Bunyan wrote his book without any intention of making it an allegory, and that it became so by accident, is just about as reasonable an idea to entertain as that the masonic legend and the emblems illustrating it were not originally designed to be what we have shown that they are—a profound and beautiful astronomical allegory.

As to the antiquity of masonry, that, we have shown, rests on the astronomical basis, and enables us to mathematically demonstrate its remote origin, independent of the uncertain and dim light of ancient history and tradition. It is true that its exact date cannot be fixed, but the proof that masonry is of great antiquity, and was founded by men of profound knowledge and exalted virtue, is conclusive: men of great learning, because their scientific knowledge lies embalmed in their work to this day; men of exalted virtue, because our ritual inculcates a code of morality never equaled or excelled until the promulgation in later times of the New Testament.

The method by which the annual progress of the sun in the zodiac is illustrated, in our explanation of the masonic allegory, also affords a key to the greater part of ancient mythology, the tales of which are founded upon the same basis, and are but so many different allegories of the same astronomical facts. When these stories were first invented by the learned, for the twofold purpose of preserving and concealing the truths of astronomy, the parallel was, of course, more perfectly preserved in each, throughout the whole narrative, than it is in the forms in which they have come down to us. Being orally transmitted, they underwent, in the lapse of long periods of time, material alterations; and particulars, not in entire harmony with the original allegory, were introduced in order to make the stories more in correspondence with the incidents of actual human life. The vulgar, who did not understand the true meaning of these astronomical parables, were most prone to make these changes. For these reasons the parallel and allegory will not be found perfect in every particular in some of them, yet in all of them enough remains of the original features to render it easy to illustrate them and their true meaning, without any material alteration of the zodiacal diagram by which we have explained the masonic legend of Hiram. It would, no doubt, be interesting to thus explain and interpret other mythological tales of antiquity, but the desired limits and special purpose of this work forbid. Having, however, pointed out the key which will unlock them all, and the method by which to conduct such an investigation, those of my readers who are curious in such matters will find their time not lost if employed in a more extended examination, from an astronomical point of view, of the poetical and wonderful adventures of the gods.

Whatever doubt may rest upon the origin of masonry, or obscurity exist as to the people among whom it first was established, it is certainly the most venerable and ancient of all existing institutions organized by man. The very obscurity as to its origin, which is lost in the dim distance of bygone ages, testifies to its real antiquity, its lodges exist in all lands, and the sound of the Worshipful Master's gavel, as he calls the brethren to order, "following the sun in its course, encircles the globe."

Its principles are as universal as its diffusion. No difference of race or color, country, clime, language, or religion, excludes any worthy and moral man from our Order. Only the atheist, the madman, or the fool, the vicious, imbecile, depraved, or degraded, are forbidden to enter our ranks, and share in all the rights, honors, and benefits of our ancient fraternity.

At our assemblies meet in harmony the Christian, the Hebrew, the Mohammedan, the Buddhist, and the Brahman, the followers of Confucius and the disciples of Zoroaster. At the masonic altar all these may offer their adoration to the same great Being in whom they all believe, the supreme great

Architect of the universe—thus presenting a sublime spectacle of the "fatherhood of God and the brotherhood of man."

The institution has frequently in times past gone through the fires of persecution, but only to rise again with its wonderful vitality renewed, and the purity and truth of its principles vindicated. At the present day it is not only one of the most popular, but also one of the most powerful and widespread, of all organizations.

While the fraternity is day by day drawing to its ranks the most intelligent and virtuous everywhere, a growing interest is also manifest among the members of the Order itself, and a disposition to inquire more fully into its origin and history, as well as to study its peculiar and beautiful system of ancient symbolical instruction. It is to be hoped that this newly awakened interest among masons will increase and bear abundant fruit, for in the ritual and emblems of our Order is a treasury of useful knowledge and sublime truth which at every step will amply reward him who diligently seeks. The subject is profound enough to enlist the highest intellect and the most accomplished scholarship. These investigations should be aided by all masons, and those engaged in them be encouraged to bring the results of their labors into the lodge-room, and communicate them for the benefit of all the brethren. Our assemblies would thus be made more interesting, and great benefits in various ways result to the fraternity.

It is the hope of the author that this work will at least aid in creating a greater interest among masons as to the history of our Order, and the true meaning of its ancient wonderful ritual. It is not expected that all readers will adopt the views of the writer; it is quite probable, on the contrary, that some will emphatically dissent from them, and, maybe, violently oppose them. But if those who disagree with the author are only induced to take a more enlarged view of the whole subject than formerly, and if in their opinion the writer is wrong in his theory as to the origin and signification of certain portions of our ritual, will themselves endeavor to discover the true solution, he will be amply satisfied with the results of his labors; for, although the author may not have discovered the truth himself, he will, perchance, thus be the cause of others doing so, and in this he will have his reward.

www.ingramcontent.com/pod-product-compliance
Lightning Source LLC
Chambersburg PA
CBHW031953040426
42448CB00006B/334